Official Statistics 4.0

Walter J. Radermacher

Official Statistics 4.0

Verified Facts for People in the 21st Century

 Springer

Walter J. Radermacher
Department of Statistical Sciences
Sapienza University of Rome
Rome, Italy

ISBN 978-3-030-31494-1 ISBN 978-3-030-31492-7 (eBook)
https://doi.org/10.1007/978-3-030-31492-7

1st edition: © Walter J. Radermacher 2019, available at IRIS (Institutional Research Information System) of Sapienza University of Rome, Italy
© Springer Nature Switzerland AG 2020

This Springer imprint is published by the registered company Springer Nature Switzerland AG
The registered company address is: Gewerbestrasse 11, 6330 Cham, Switzerland

Preface

Statistics in the form of key figures, graphs and rankings play an increasingly important role in everyday life today. One characteristic of good governance is that political decisions are evidence-based. Facts and figures, with their scientific and technical nature, appear to be outside the political realm and thus immune to any infection by political interests. At the same time, it is precisely this form of governance, based on expert knowledge and facts, that has recently developed into a deep-seated mistrust, which has led to an influx of those forces in politics that consciously and deliberately cast doubt on the existence of neutral facts. If, in this way, everything is put into perspective and citizens' confidence in institutions and numbers is reversed, then who can be trusted?

The book shows a way out of this dilemma by taking facts for what they are, produced information. This means that it is not about any absolute truth, but about high-quality information, which is well worth trusting. What information quality is, how this quality can be designed, produced and certified and how users can assure themselves of this quality is the focus of this publication, which draws conclusions from the history of the first three chapters of 200 years of official statistics for the upcoming and necessary adjustments in the future. Understanding the driving forces of science, statistics and society and their interplay is of pivotal importance for this objective.

Rome, Italy   Walter J. Radermacher

Contents

Abbreviations

AI	Artificial intelligence
AMStat	American Statistical Association
B2G	Business-to-government
CBS	Central Bureau of Statistics
COs	Citizens' Observatories
CPI	Consumer Price Index
CSO	Central Statistical Office of Ireland
DGINS	Conference of the Directors General of the National Statistical Institutes
DMC	Domestic Material Consumption
EC	European Community
ECB	European Central Bank
EDP	Excessive Deficit Procedure
EEA/EFTA	European Economic Area/European Free Trade Association
EFQM	European Foundation for Quality Management
EGR	EuroGroups Register
ELSTAT	Hellenic Statistical Authority
EMOS	European Master in Official Statistics
ES Code of Practice	European Statistics Code of Practice
ESA	European System of Accounts
ESBRs	European System of interoperable Business Registers
ESS	European Statistical System
EU	European Union
Eurostat	European Statistical Office
FENStatS	Federation of European National Statistical Societies
G2B	Government-to-business
GDP	Gross Domestic Product
GSBPM	Generic Statistical Business Process Model
HICP	Harmonised Index of Consumer Prices

HLEG	High-level expert group of experts
ICW	Income, Consumption and Wealth
IMF	International Monetary Fund
Intrastat	Statistics on the trade in goods between countries of the European Union
IoT	Internet of Things
ISI	International Statistical Institute
IT	Information Technology
LEG	European Statistical System Leadership Expert Group
LFS	Labour Force Survey
MDGs	UN Millenium Development Goals
NDP	Net Domestic Product
NGOs	Non-Governmental Organisations
NUTS	Nomenclature of Territorial Units for Statistics
OECD	Organisation for Economic Co-operation and Development
PDSA	Plan-Do-Study-Act
QAF	Quality Assurance Framework
RatSWD	German Data Forum
SD	Sustainable Development
SDG	UN Sustainable Development Goals
SDI	Sustainable Development Indicators
SEEA	System of Environmental-Economic Accounting
SNA	System of National Accounts
TIVA	Trade-in-value-added
TMC	Total Material Consumption
TQM	Total Quality Management
UK	United Kingdom
UN	United Nations
UNECE	United Nations Economic Commission for Europe

Chapter 1
Official Statistics—An Introduction

*To measure for public purposes is rarely so simple as to apply a
meter stick casually to an object.*
Porter (1995, p. 28)

*In Wirtschaft und Gesellschaft bestimmt das von historischen,
institutionellen und kulturellen Rahmenbedingungen abhängige,
an Werten und Normen orientierte, vielfach interessengeleitete
Verhalten der Menschen so weitgehend das Geschehen, dass
schon eine sinnvolle Begriffsbildung und damit auch die
Datenerhebung einen ganz eigenen, geradezu kulturorientierten
Zugang erfordern. Das ist das Adäquationsproblem.*[*]
Grohmann (2012, p. 59)

The term 'statistics' is used differently; it can refer to a science, a certain kind of information or institutions.

Essentially, **statistics** is the **science** of learning from data. Certainly, it is a modern technology that is part of the standards of today's information age and society and is used in a wide array of fields. The history of statistics goes back a long way, accompanying historical eras, technical developments and political turning points just as the census in year zero[1] (Champkin 2014).

Statistics is a method that can reduce complexity, separate signals from noise and distinguish significant phenomena from random dispersion. The **statistical results** of this method are used for all conceivable information and decision-making processes. Whether statistics help us better understand the world around us and whether they actually improve decisions (and therefore our lives) are not only questions of scientific methodology. The decisive factor here is whether statistics, like a language, are understood by those for whom the information is relevant.

---

[*]In business and society, people's behaviour, which depends on historical, institutional and cultural conditions, is oriented towards values and norms and is often guided by interests and determines what happens to such an extent that even the formation of meaningful concepts and thus the collection of data require a very individual, almost culture-oriented approach. This is the problem of adequacy.

[1]Luke 2:1: *"In those days Caesar Augustus issued a decree that a census should be taken of the entire Roman world."*

© Springer Nature Switzerland AG 2020
W. J. Radermacher, *Official Statistics 4.0*,
https://doi.org/10.1007/978-3-030-31492-7_1

Statistical institutions are the producers of statistics. Using scientific statistical methods, data is collected and existing data is processed in order to calculate condensed information (i.e. facts), which is made available to the general public in different forms, such as statistical aggregates, graphics, maps, accounts or indicators. Statistical offices usually belong to the public administration, at state, international, regional or local level.

This work will be concerned neither with statistics in general nor with the history of theoretical statistics. Rather, the goal is to describe the status quo for a particular area of application, namely 'official statistics', based on an analysis of its historical genesis in order then to deploy strategic lines of development for the near future of this particular domain.

Central to this work is the quality of statistical information. Statistics can only develop a positive enlightenment effect on the condition that their quality is trusted. To ensure long-term trust in statistics, it is necessary to deal with questions of knowledge, quantification and the function of facts in the social debate. How can we know that we know what we know (or do not know)? The more concrete an answer that can be given to such questions, the more possible it will be to protect statistics against inappropriate expectations and to address false criticism.

When one uses the term 'official statistics', one deals with the problem that again different meanings are possible, namely the institution (the statistical office), the results (statistical information) and, of course, the processes (the surveys). As we will see below, such a still very vague interpretation is actually not entirely wrong. To define 'official statistics' means to commit oneself to all three questions: who? what? and how?

But of course, one associates with the notion of official statistics first that it deals with social and economic issues, and more recently also with ecological ones. How many people live in a country, how much is produced, what about work, health and education? We encounter these and related topics daily in the media, in political discussions and decisions. From them, we expect a solid quality; we must trust them.

In fact, official statistics is a representative of the "*Statistical Mind in Modern Society*" (Stamhuis et al. 2008; van Maarseveen et al. 2008), closely related to social progress and scientific work. In this respect, it is not surprising that the interrelationships between statistics, science and society are reflected in a historical development characterised by manifold turns, by steady sections, alternating with periods of greater and more rapid change.

From the beginning of the nineteenth century, in the course of the emerging nation states and in parallel with the Industrial Revolution, statistics experienced a **first phase** of growth, methodological development and various applications. Statistics as a science, as a statistical result and as an institution fertilised each other in their development, although there were consistently disagreements between different schools of thought, especially between the representatives of empirical, comparative statistics on the one hand and of a '*stochastic style of reasoning*' (Desrosières 2008a, p. 311) on the other.

In the twentieth century, three methodological and technical innovations have changed the world of official statistics: "*sampling surveys, national accounts and computers*" (Desrosières 2008a, p. 320). After a first era of official statistics in the nineteenth century, a **second phase** of the prosperity of statistics followed, mainly initiated by major scientific innovations particularly in inferential statistics, but also closely connected with the political conditions, the crises and the attempts to solve them, for example, by the development of macroeconomic statistics.

In a **third phase**, which began in the late 1970s, the computer moved into the spaces and processes of statistics, which opened up completely new possibilities. The amount of data and the variety of its processing tools in all areas of life, commerce, administration, politics has exploded since then. In this third era, under these conditions, official statistics were fundamentally reformed by switching from tailor-made to industrial production processes.

At the end of the third era, we are currently in a transition to a **fourth phase** in which the digitisation of all areas of life will continue at high speed. The handling of 'Big Data' will dominate the near future of official statistics as the question of register data has done in recent years. In addition, the effects of globalisation will increasingly demand political responses, which will then be directly linked to a new need for differentiated statistics (Fig. 1.1).

'Official statistics' is one practical application of the 'quantification as a social technology' (Porter 1995) belonging to those with the longest history.[2] Since the beginning of the nineteenth century (von Schlözer 1804), official statistics—as a child of the enlightenment—have grown and developed side by side with the different forms of the (modern) state.

Desrosières (1998) uses the term 'mutual co-construction' for three interlinked phenomena: (a) a theory of the state (economy); (b) interventions of the state (policies); and (c) quantification of 'variables' specifically targeted by policy measures (statistics).

Generally, the question 'what is official statistics?' is not taken very seriously. It is only inadequately answered or often even answered with a certain irony: 'official statistics is statistics produced by offices'. In any case, little importance is attributed to the question by the academic representatives of the scientific discipline of statistics since they regard official statistics as nothing other than the application of methodology in one field of practice, as well as others (e.g. medicine or industry).

In this work, however, the exclusive focus is placed on what official statistics is, how it came into being, what significance it has for sociopolitical processes (and vice versa) and where the developments will lead in the near future. It is therefore imperative that we examine this subject of consideration more closely and at the same time describe it to the extent that this is possible with an abstract definition.

In an approximation, official statistics can be defined by using three questions (Eurostat 2016):

- Who? Normally, official statistics are produced and provided by statistical offices, i.e. public administrations.

---

[2]For a comprehensive overview of the history of official statistics, reference is made in particular to the works of Desrosières, Porter and Klep, Stamhuis et al. (Desrosières 2008b, c; Porter 1986, 1995, 2004; Klep and Stamhuis 2002; Stamhuis et al. 2008; van Maarseveen et al. 2008).

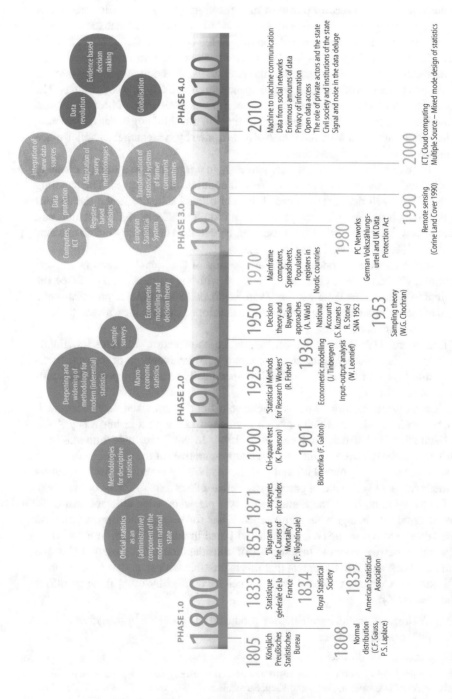

Fig. 1.1 Timeline of official statistics

- What? Statistical work programmes and priorities are prepared according to public sector standards (i.e. participation of civil society) with the final decisions partly taken in legislative procedures.
- How? Statistical methodologies are nowadays subject of international cooperation and manifested in statistical standards; high-level quality is assured through management systems and ethical codes.

Due to this somewhat more complicated rapprochement, it is already clear that official statistics are characterised above all by their role and function in the process of forming opinions and deciding in societies. We will first examine this question in more detail, and in doing so, we will approach the main field of inquiry itself, namely the interaction between official statistics and society.

The need for statistics has never been so obvious (Radermacher 2012b, 2016). Data requirements cover a wide range of aspects of society, including relatively new areas such as quality of life, environmental aspects or the economy 4.0. The financial and economic crisis since 2007 has led to stronger economic governance in the European Union (EU), which in turn has led to a greater need for reliable, trustworthy statistics.

Official statistics play a fundamental role in modern societies: they are an essential basis for policies, they support business decisions, and they allow citizens to evaluate the progress made. But the power of statistical knowledge also poses dangers (Fukuda-Parr 2017). From a cognitive tool that can emancipate and promote participation, it can transform itself into a true technocratic tyrant, to varying degrees, behind evidence-based decision-making and mainstream management ideologies[3] (Davis et al. 2012a, b; Sangolt 2010; Brown 2015).

In principle, official statistics enable anyone to observe and assess social, economic and ecological phenomena. They provide evidence for the formation of opinion(s), but they are neither an end in themselves nor a substitute for decisions. They need to clarify the availability of alternatives and facilitate their selection, but without taking sides themselves. They are a political element, not a politics in themselves (Turnpenny et al. 2015). But the temptation of the power in them is strong and their attraction almost magnetic. Therefore, official statistics should not be reserved for use by technical experts. Statisticians need to engage with the public and work intensively and regularly with different users and stakeholders, whether public or private, journalists, researchers or citizens. The goal is to better understand their needs (as users of statistics) and their limitations (as sources of statistics) in order to provide them with adequate information. To do this, statisticians must actively seek to create a positive data culture by becoming more flexible and reactive to ensure that official statistics are understood well (Radermacher 2012a). With the intelligent tools available today, such as interactive graphics, the contents of the partially abstract information provided by official statistics can be communicated much better. Of course, it is very important to strike a balance between the dissemination of understandable messages and a strict focus on technical precision, between excessive simplification and unnecessary complexity, between vulgarisation and overly scientific methods

---

[3] See in particular Sect. 3.3.

and outcomes. Likewise, the boundaries between objective, quantifiable conditions and subjective impressions must be clearly demonstrated.

This short summary explains the special role and function of official statistics for policy-making.[4] Against this background, it should again be emphasised that the mandate of official statistics cannot be characterised solely by the fact that statistical methods of the social sciences are used. Rather, a wider description is needed to cope with the diversity of official statistics[5] in order to cover the following core components:

Official statistics represent a public information infrastructure, a system of informational products that meets a variety of needs, including scientific quality, transparency and excellence.

Another element of the 'Markenkern' (the brand essence) refers to the subjects of observation and accordingly the 'variables' (such as GDP, employment, income or inflation) that are closely related to policy-making and society, both in concepts and in reality, reflecting highly aggregated artefacts.[6] These variables are designed in a separate process (adequacy) before they can be measured (Grohmann 1985; Radermacher 1992, 2017). The process of design is primarily based on an optimal use of the currently available statistical methods. Nonetheless, these variables contain conventions and decisions that, to justify their 'authority', must be embedded in democratic and participatory processes. The set of statistical standards (including the statistical programme) is a service provided by official statistics to citizens, entrepreneurs and politicians. It is a valuable asset for official statistics.

Official statistics have very different facets and dimensions, which are related to neighbouring fields of science, the judiciary, industry, design, civil service and the media. Above all, however, official statistics are closely tied to the political position and role of the state as an institution and in particular that of the 'modern' nation state, as it emerged in the nineteenth century. Only when all these roles and dimensions are seen together and understood as a bundle of functions, does one do justice to official statistics. A focus on the scientific function alone is just as irrelevant as an isolated emphasis on the state-political position. As the history of official statistics has shown, progress is blocked (von der Lippe 2012) when the triangle of driving forces (science, statistics, society) is not understood as an 'eternal golden braid' (Hofstadter 1979).

This work will begin in Chap. 2 by describing official statistics, at the current stage of development observable in Western industrial nations and in particular the EU. The stamp 'Official Statistics 3.0' is intended to clarify parallels and associations with the phases of industrialisation (Fig. 1.2).

---

[4]This role is described in detail in Porter (1995) and Desrosières (1998).

[5]Desrosières explains: *"Almost since its origin statistics has had two different but intertwined meanings: on the one hand denoting quantitative information, collected by the state ... and, on the other, mathematical techniques for treatment of and argument over facts based on large numbers..."* (Desrosières 2010, p. 112).

[6]'Variables' uses the terminology of Desrosières, which distinguishes between *"making numbers"* (or "data") and *"making variables"* (or statistical "constructs") and the embedding of variables in more complex models (such as National Accounts) (Desrosières 2010, p. 114).

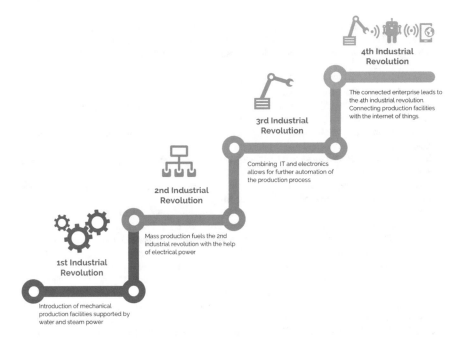

Fig. 1.2 Industrial mechanisation. *Source* https://www.roboyo.de/en/in-focus/industry-4-0/ based on Plattform Industrie 4.0 (2013); published with permission

At the end of a period of automation, official statistics today are considerably more efficient, faster and better in quality compared to the 1980s. But what has now begun with a comprehensive digitisation of our world demands in many ways a more radical and innovative strategy. In addition, globalisation calls for new answers that cannot always be achieved with continuous development, but that will be accompanied by disruptions and more radical changes.

In preparation for deriving conclusions and strategies in Chap. 4, it is helpful, indeed necessary, to take in Chap. 3 a closer look at the driving forces: the scientific background of official statistics as well as episodes from the history of the past 200 years, insofar as they are relevant to the understanding of Statistics 4.0.

References

Brown, W. 2015. *Undoing the Demos: Neoliberalism's Stealth Revolution*. Cambridge, MA: MIT Press.

Champkin, Julian. 2014. 'The Timeline of Statistics', *Significance*, January 24, 2014.

Davis, Kevin E., Angelina Fisher, Bendict Kingsburry, and Sally Engle Merry. 2012a. *Governance by Indicators—Global Power Through Quantification and Rankings*. Oxford: Oxford University Press.

Davis, Kevin E., Benedict Kingsburry, and Sally Engle Merry. 2012b. Introduction: Global Governance by Indicators. In *Governance by Indicators: Global Power Through Quantification and Rankings*, ed. Kevin E. Davis, Benedict Kingsburry, and Sally Engle Merry. Oxford: Oxford University Press.

Desrosières, Alain. 1998. *The Politics of Large Numbers—A History of Statistical Reasoning*. Cambridge, MA: Harvard University Press.

Desrosières, Alain. 2008a. Commensuration and Probabilism: Two Kinds of Controversies About Statistics (1850–1940). In *The Statistical Mind in Modern Society. The Netherlands 1850–1940—Volume II: Statistics and Scientific Work*, ed. I.H. Stamhuis, P.M.M. Klep, and J.G.S.J. van Maarseveen. Aksant: Amsterdam.

Desrosières, Alain. 2008b. *Gouverner par les nombres: L'argument statistique II*. Paris: Les Presses de l'École des Mines.

Desrosières, Alain. 2008c. *Pour une sociologie historique de la quantification: L'argument statistique I*. Paris: Les Presses de l'École des Mines.

Desrosières, Alain. 2010. A Politics of Knowledge-Tools—The Case of Statistics. In *Between Enlightenment and Disaster*, ed. Linda Sangolt. Brussels: P.I.E. Peter Lang.

Eurostat. 2016. *Quality Declaration of the European Statistical System*. Luxembourg: Eurostat.

Fukuda-Parr, Sakiko. 2017. *United Nations High Level Political Forum Opening Panel, 10 July 2017—Statement by Sakiko Fukuda-Parr*. New York: United Nations.

Grohmann, Heinz. 1985. Vom theoretischen Konstrukt zum statistischen Begriff - Das Adäquationsproblem. *Allgemeines Statistisches Archiv* 69: 1–15.

Grohmann, Heinz. 2012. Ein Blick auf Geschichte und Bedeutung der Deutschen Statistischen Gesellschaft - Ansprache anlässlich des 100-jährigen Bestehens der Deutschen Statistischen Gesellschaft am 21. September 2011 in Leipzig. *AStA Wirtsch Sozialstat Arch* 2012: 57–60.

Hofstadter, Douglas R. 1979. *Gödel, Escher, Bach—An Eternal Golden Braid*. New York: Basic Books.

Klep, P.M.M., and I.H. Stamhuis. 2002. *The Statistical Mind in a Pre-statistical Era: The Netherlands 1750–1850*. Aksant.

Plattform Industrie 4.0. 2013. Umsetzungsempfehlungen für das Zukunftsprojekt Industrie 4.0 - Abschlussbericht des Arbeitskreises Industrie 4.0. Frankfurt/Main: Plattform Industrie 4.0.

Porter, Theodore M. 1986. *The Rise of Statistical Thinking: 1820–1900*. Princeton, Guildford: Princeton University Press.

Porter, Theodore M. 1995. *Trust in Numbers: The Pursuit of Objectivity in Science and Public Life*. Princeton, N.J., Chichester: Princeton University Press.

Porter, Theodore M. 2004. *Karl Pearson: The Scientific Life in a Statistical Age*. Princeton, NJ, Oxford: Princeton University Press.

Radermacher, Walter. 1992. Methoden und Möglichkeiten der Qualitätsbeurteilung von statistischen Informationen aus der Fernerkundung/Methods and Possibilities of Assessing the Quality of Statistical Data of Remote Sensing. *Jahrbücher für Nationalökonomie und Statistik* 169–179.

Radermacher, Walter. 2012a. Aus Zahlen lernen: "Die Statistikämter sollen sich als Institutionen politischer Bildung verstehen" Interview. *Weiterbildung* 4, August/September 2012.

Radermacher, Walter J. 2012b. Zahlen zählen - Gedanken zur Zukunft der amtlichen Statistik in Europa. *AStA Wirtsch Sozialstat Arch* 2012: 285–298.

Radermacher, Walter J. 2016. Les Statistiques Comptent - éviter le syndrome du lampadaire et avoir un citoyen européen éclairé. *La Revue France Forum*, no. 61.

Radermacher, W.J. 2017. Governance der amtlichen Statistik. *AStA Wirtsch Sozialstat Arch* 11: 65–81. https://doi.org/10.1007/s11943-017-0207-7.

Sangolt, Linda. 2010. A Century of Quantification and "Cold Calculation." Trends in the Pursuit of Efficiency, Growth and Pre-eminence. In *Between Enlightenment and Disaster—Dimensions of the Political Use of Knowledge*, ed. Linda Sangolt. Brussels: P.I.E. Peter Lang.

Stamhuis, Ida H., Paul M.M. Klep, and Jacques G.S.J. van Maarseveen (eds.). 2008. *Statistics and Scientific Work*. Amsterdam: Aksant.

Turnpenny, John R., Andrew J. Jordan, David Benson, and Tim Rayner. 2015. The Tools of Policy Formulation: An Introduction. In *The Tools of Policy Formulation*, ed. Andrew J. Jordan and John R. Turnpenny. Cheltenham: Edward Elgar.

van Maarseveen, Jacques G.S.J., Paul M.M. Klep, and Ida H. Stamhuis (eds.). 2008. *Official Statistics, Social Progress and Modern Enterprise*. Amsterdam: Aksant.

von der Lippe, Peter. 2012. *Was hat uns die "Frankfurter Schule" der Statistik gebracht? - Darstellung einer deutschen Fehlentwicklung am Beispiel der Indextheorie von Paul Flaskämper*. Duisburg: Institut für Betriebswirtschaft und Volkswirtschaft (IBES) - Universität Duisburg-Essen.

von Schlözer, August Ludwig. 1804. *Theorie der Statistik. Nebst Ideen Ueber Das Studium Der Politik Überhaupt*. Vandenhoek und Ruprecht: Goettingen.

Chapter 2
Official Statistics—Public Informational Infrastructure

This chapter is about the 'making of' official statistics. The processes, structures and actors that are crucial for the high quality are to be presented. Official statistics are understood as industry that produces information. Consequently, in the presentation, the terms and concepts of modern management are used throughout.

The first section starts with the business model of statistics with its dimensions of the processes ('how'), the products ('what') and the producers ('who'). It then deals with important overarching topics, such as quality management, national and international statistics and statistical confidentiality. With a look at the recent modernisation of the business model, the current status of Statistics 3.0 is summarised.

2.1 The Business Model of Official Statistics

2.1.1 Core Aspects

Many of the essential definitions and foundations of official statistics can be found in the statutory provisions of Regulation 223 on European statistics.[1] These represent an agreement of the partners cooperating in the European Statistical System (EU as well as Switzerland and EEA/EFTA, currently a total of 32 states), but which also applies in other European countries (e.g. candidates for EU accession).[2]

The regulation

> establishes a legal framework for the development, production and dissemination of European statistics. (Art 1):

---

[1] See more detailed comments on the legal provisions in Radermacher/Bischoff "*Article 338*" (Radermacher and Bischoff 2018 forthcoming).

[2] The European Free Trade Association (http://www.efta.int/about-efta/european-free-trade-association) and the European Economic Area (http://www.efta.int/eea) and for statistics http://www.efta.int/statistics.

© Springer Nature Switzerland AG 2020
W. J. Radermacher, *Official Statistics 4.0*,
https://doi.org/10.1007/978-3-030-31492-7_2

The development, production and dissemination of European statistics shall be governed by the following statistical principles (Art 2):

- professional independence
- impartiality
- objectivity
- reliability
- statistical confidentiality
- cost effectiveness

The statistical principles set out in this paragraph are further elaborated in the European Statistics Code of Practice. The development, production and dissemination of European statistics shall take into account international recommendations and best practice.

The following definitions shall apply (Art 3):

- 'statistics' means quantitative and qualitative, aggregated and representative information characterising a collective phenomenon in a considered population;
- 'development' means the activities aiming at setting up, strengthening and improving the statistical methods, standards and procedures used for the production and dissemination of statistics as well as at designing new statistics and indicators;
- 'production' means all the activities related to the collection, storage, processing, and analysis necessary for compiling statistics;
- 'dissemination' means the activity of making statistics and statistical analysis accessible to users;
- 'data collection' means surveys and all other methods of deriving information from different sources, including administrative sources;
- 'statistical unit' means the basic observation unit, namely a natural person, a household, an economic operator and other undertakings, referred to by the data;
- 'confidential data' means data which allow statistical units to be identified, either directly or indirectly, thereby disclosing individual information. To determine whether a statistical unit is identifiable, account shall be taken of all relevant means that might reasonably be used by a third party to identify the statistical unit. (European Union 2015)

The following sections build on these foundations; they are interpreted and further elaborated.

2.1.2 Knowledge Generation

A simplified circular process chart describing the interaction between users and producers of information should help us to understand the main phases in the production and the use of statistical information:

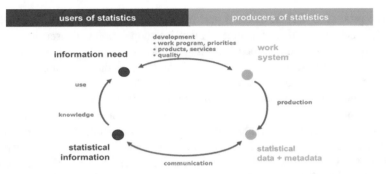

Fig. 2.1 Knowledge generation and statistical production [See earlier versions in Blanc et al. (2001) and Radermacher et al. (2004)]

The key-processes within the **production** sphere in Fig. 2.1 are[3]

D: development and design

- Input: information requests and needs for statistical information expressed in qualitative form (language)
- Output: a work system that contains the necessary statistical specifications (variables, methodology, standards, sampling design, etc.) and concrete prescriptions with regard to the entire work programme and individual production lines.

P: production

- Input: specifications of the work system
- Output: statistical data and metadata.

C: communication/dissemination

- Input: statistical data and metadata
- Output: statistical information.

In addition, it is essential to include explicitly the following process on the **user** side.

U: creation of knowledge and application

- Input: statistical information
- Output: quantitative response to the qualitative information request.

The ultimate goal of statistical evidence is to contribute to better informed decisions of all kind and for all types of users, which can only be achieved when all four processes are considered and integrated in a comprehensive conceptual approach. Each of them should contribute to excellent information quality. Each of them can of course also fail and contribute to errors, misunderstandings and underperformance.

---

[3]This circular flow chart corresponds to widely accepted standards on the producer's side, such as the Generic Statistical Business Process Model (GSBPM) (UNECE 2013) or the Generic Statistical Information Model (GSIM) (UNECE 2017).

The process D has an external part (dialogue with users) and an internal part (development and testing of methods). Intensive cooperation with users is crucial for the adequacy of the entire process chain that follows.

During the production process P, the methods agreed in the preceding development phase are implemented. It is relatively straightforward to measure the quality of this process and its sub-processes against these predefined norms.

Communication processes C represent the other end of the user interaction. They can also be grouped into an internal part (preparing the results from the production process for different channels, access points, etc.) and an external part (interaction with users in all formats and through all channels). The internal part also belongs to the set of predefined methods and is in that way similar and closely linked to production.

The processes of application and use U are not under any kind of control or influence by statisticians. It is, however, obvious that users might not be sufficiently prepared or trained to interpret and use statistics in the best possible manner. Statistical literacy is therefore an area of interest also for statistical producers. Furthermore, statisticians should carefully observe cases of wrong interpretation and they must protect their information against misuse.

2.1.3   The Process Model, Business Architecture

The flow-model of knowledge generation and statistical production process (Sect. 2.1.1) can be further used and elaborated for the creation of a generic process model of official statistics, using the format of an input-output flowchart (Fig. 2.2).

At the centre are individual production processes of specific statistics, starting with a survey the results of which are condensed in data processing into information that is analysed and published, thus, finalising the process. Close to these core processes were also the support processes and corresponding internal services (publication, IT, etc.). A highly branched organisation of these individual processes in isolated 'silos'

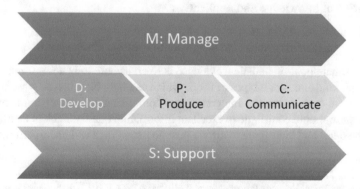

Fig. 2.2 Main processes in official statistics

was the historically grown one-to-one relationship between the producers and users of the individual statistical areas.[4] Thus, the agricultural statistics unit produced as closely as possible what was desired by the Ministry of Agriculture; similar in economics, health, energy, etc. In total, this resulted in more than 200 parallel processes: a veritable spaghetti bowl.

Statistics in this logic was tailor-made and crafted for the needs of a particular customer (or customer group). For each of these areas, therefore, more or less the entire procedure schematised in the GSBPM (UNECE 2013) has been completed separately and without feedback from similar areas. In such an understanding of the manner in which statistics are produced, there are, thus, primarily individual production lines which are only weakly and insignificantly connected to each other. Such strands can therefore be organised, opened, closed and financed without any major impact on other areas.

Information technology has dramatically improved the possibilities of official statistics over the past four decades. However, these new possibilities have ultimately contributed to the fact that the already fragmented organisation disintegrated even more into heterogeneous and inefficient parts. While mainframe information technology was very centralised in the 1970s and 1980s, the introduction of personal computers also resulted in a wave of decentralisation in the 1990s and 2000s.

Not least because of the reduced budgets and resources, this form of official statistics was no longer possible, at least since the beginning of the 2000s. The isolated process organisations lacked efficiency and consistency. Parallel and non-coordinated areas of production have been targeted by reforms and modernisation projects (Eurostat 2009). Generally, this modernisation aims at substituting the stovepiped way[5] of working by a new form, i.e. a new business model, which can be summarised as 'Multiple Source—Mixed Mode Design' at the data-input side, with a 'multipurpose design' at the information-output side and with a modularisation of exchangeable process elements[6] within a standardised business architecture at the centre of the statistical factory. This will be discussed in more detail in Sect. 2.6.

In this context, statistical offices considered ways of making production more uniform so as to be more efficient and effective. The result of these considerations led to a kind of industrialisation of the processes with the typical components, i.e. standardisation (of methods, IT applications, etc.), centralisation of common components (IT, auxiliary services, etc.) and, last but not least, the introduction of an overarching business architecture as an ordering system. The flowchart in Fig. 2.3 explains this architecture in a very simplified and graphical manner.

Such a business model of the 'factory' is still relatively new and does not necessarily meet with the approval and sympathy of those who work in this institution.

---

[4]This stovepipe approach in the organisation of work is further strengthened if the financing of statistical branches is separated in different budget lines.

[5]This way of producing statistics in parallel but only weakly coordinated processes could be called a 'vertical organisation'.

[6]This way of producing statistics with exchangeable modules could be called a 'horizontal organisation'.

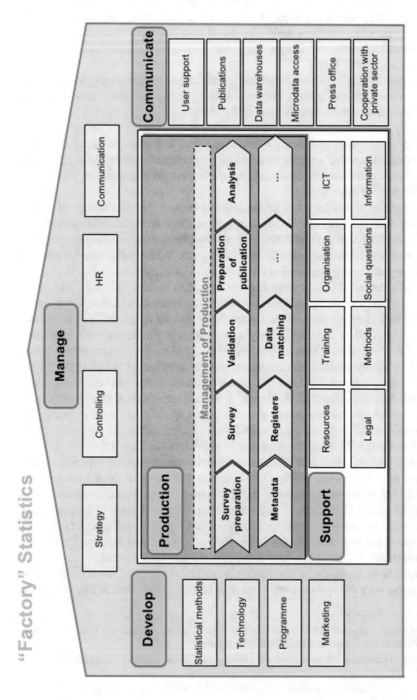

Fig. 2.3 Business architecture of official statistics (Radermacher 2011)

Centralisation and standardisation are perceived as a loss of self-responsibility, and the replacement of a craft by an industrial model is sensitive to the professional self-understanding of statisticians. Nevertheless, there is no way around this approach. Too powerful and urgent are the constraints of the general political situation and the dynamics of developments in information and communication technology.

2.1.4 Modes of Data Collection

It is one of the myths about official statistics that it is exclusively based on self-collected data. While this is true for large areas of economic and social statistics, in other areas, such as demography, health care statistics or education statistics, however, existing data sources are being evaluated. This has never been different; rather, it has been a standard approach, at least in the early days, before high-quality sample surveys were methodically enabled. Primary sources of data collection under the control and responsibility of the statistical office make up not more than half of the processes, while others evaluate existing (secondary) data sources, including registers. Even if this picture refers to the situation in Germany in 2008 (see Fig. 2.4), it is representative of the fact that official statistics is by far not a data collection engine.

However, the data from different sources were generally not merged and used together to generate information. Rather, it was left to the users, in case of parallel running statistics with different origins, to make the right choice for themselves. Merging data from different sources requires rules (and algorithms) that allow synthesis to be transparent and not arbitrary; this would violate basic principles. For a

Data collection in official statistics

	Primary sources	Secondary sources	Statistical registers
Definition	Data from surveys which have been designed and carried out exclusively for statistical purposes	Data which were collected for non-statistical purposes (e.g. by public administration)	Combination of (micro) data from one or more administrative registers and (additional) surveys
Data source	Direct survey of the survey units	Indirect data collection via public administration and other institutions	Transfer of data from administrative registers (plus additional primary surveys)
Data collection method	Mail, web, face-to-face, or telephone interview based on a questionnaire	Electronic (or mail) data transmission	Electronic transmission of micro data from secondary (and primary) sources; combination of various sources
Germany:	ca. 175 statistics	ca. 140 statistics	7 registers

Fig. 2.4 Primary and secondary modes of data collection—Germany 2008 (Radermacher 2008)

long time, it was not considered to be the job of official statistics to do this 'blending'. Instead, more restraint was maintained, and unprocessed results were provided. Last but not least, producers and users at that time agreed in the conviction that survey data was seen as superior to administrative data sources. Only in National Accounts was it considered inevitable and opportune to distil the best possible information from multiple data sources, to close data gaps with estimates in order to arrive at a complete and consistent picture.

This has changed.

The primacy of survey data over existing data sources was unsustainable for many reasons. In the end, it was a mixture of increasing availability of data in (administrative as well as statistical) registers, the potential of new IT (online transfer of data), cost pressures, and dissatisfaction of respondents with statistical burdens that reversed the prioritisation to its opposite. According to a modern prioritisation, it is appropriate and legitimate to collect data if and only if these data cannot already be obtained from existing sources of satisfactory quality. This opens a door to a completely different business model with fundamental changes in the tasks of a statistical office, with new components in the methodological toolbox (e.g. record linkage), with adaptations of the statistical governance, e.g., the creation of legal-administrative conditions (access to some sensitive administrative data[7]) as well as changes in quality management and in communication to users.

One might have the impression that the melange between survey and administrative data is nothing but replacing an item in a questionnaire by a similar piece of data from a register. This impression has been shown to be much too simplistic and not realistic.[8] Instead, the entire design of one statistical process has to be reviewed and (quite often) revised. It is a long way from the classical design, where a traditional ('knock on the door') census survey every ten years was alternating with yearly interpolations of the population from administrative registers to a modern (fully integrated) design, where a regular matching of administrative and (sample) survey data ensures the best possible capturing of high dynamics in population changes on a continuous (yearly) bases, delivering a completely different mix of quality features (improvements in timeliness, coherence at the expense of accuracy in the traditional census years). It is hardly possible to overestimate the difficulties of change management in the transition from the traditional to the new design. Not all users are winners in such a change and not all producers welcome the changed production and their products.

In particular, the Nordic countries have re-engineered their statistical systems by shifting them entirely to the prioritised use of registers (see Fig. 2.5).

---

[7]See for example Bundesstatistikgesetz, § 5a Nutzung von Verwaltungsdaten (https://www. destatis.de/DE/Methoden/Rechtsgrundlagen/Statistikbereiche/Inhalte/010_BStatG.pdf?__blob= publicationFile).

[8]See for example "*Good practices when combining some selected administrative sources*" https:// ec.europa.eu/eurostat/cros/system/files/good_practices_administrative_data.pdf.

Type of register	Denmark		Finland		Norway		Sweden	
	Established	*First used in census*	*Established*	*First used in census*	*Established*	*First used in census*	*Established*	*First used in Census*
Central Population Register	1968	1981	1969	1970	1964	1970	1967	1975
Business Register	1975	1981	1975	1980	1965	1980	1963	1975
Dwellings	1977	1981	1980	1985	2001	*2011*	*2008?*	*2011?*
Housing conditions	1977	1981	1980	1985	2001	*2011*	*2008?*	*2011?*
Education	1971	1981	1970	1975	1970	1980	1985	1990
Employment	1979	1981	1987	1990	1978	2001	1985	1985
Family	1968	1981	1978	1980	1964	1980	1960	1975
Household^a	1968	1981	1970	1975	2001	*2011*	*2011?*	*2011?*
Income	1970	1981	1969	1970	1967	1980	1968	1975
Totally register-based census		1981		1990		*2011*		*2011?*

Fig. 2.5 Year of establishing registers in population censuses [From Register-based statistics in the Nordic countries, by UNECE Statistical Division, © 2007 United Nations. Reprinted with the permission of the United Nations. UNECE (2007, p. 5)]

This Nordic way cannot be followed in the same manner by every country. The legal conditions of access to individual data corresponding to administrative practices and political as well as cultural conditions (presence of high-quality registers, trust of citizens in government institutions, etc.) are too different. Nonetheless, the fundamental approach is widely used in the reality of official statistics nowadays. The trend towards population censuses, which are created entirely or partially from register data, illustrates this statement.[9]

In this respect, the consideration and interaction of 'Big Data' is nothing fundamentally different; the paradigm shift has already taken place. However, the task of statistics is further complicated because the possibility to influence the nature and structure of this external data continues to diminish (more precisely: no longer exists), but at the same time the general pressure and the expectation that it has to be used has increased immensely.

---

[9]See for example Darabi (2017) or Kyi et al. (2012).

2.1.5   The Portfolio of Products (And Services)

2.1.5.1   Statistical Products (And Services)

As a starting point for the consideration of statistical products, the definition in EU regulation 223 is used again: *"statistics' means quantitative and qualitative, aggregated and representative information characterising a collective phenomenon in a considered population'* (European Union 2015: Art. 3).

For an understanding of the functioning and internal organisation of official statistics, it is necessary to arrange different levels and types of statistical information according to their degree of aggregation and their quality profile.

At European level,[10] the following types of statistical products are distinguished:

- Data: information collected by statistical authorities, via traditional statistical activities (sample surveys, censuses, etc.)/data from other sources, re-used for statistical purposes. This information is tailored to serve needs in specific policy areas, e.g. the labour market, migration or agriculture. The term also includes data collected for administrative purposes but used by statistical authorities for statistical purposes (usually referred to as data from administrative sources).

- Accounting systems: coherent and integrated accounts, balance sheets and tables based on a set of internationally agreed rules. An accounting framework ensures a high profile of consistency and comparability; statistical data can be compiled and presented in a format that is designed for the purposes of analysis and policy-making.

- Indicators: an indicator is a summary measure related to a key issue or phenomenon and derived from a series of observed facts. Indicators can be used to reveal relative positions or show positive or negative change. Indicators are usually a direct input into EU and global policies. In strategic policy fields they are important for setting targets and monitoring their achievement.

This view can be condensed to an information pyramid (Fig. 2.6).

Primarily, this presentation relies on a distinction of different aggregation levels, i.e. a level with many details (i.e. micro) for the basic statistics and a level with more abstract aggregates and models (i.e. macro) for accounts and indicators.

Furthermore, basic statistics and accounts are characterised as multipurpose,[11] while indicators are closely tied to a specific use and determination.

- 'Multipurpose' makes clear that such statistical information has the character of an infrastructure designed for wide and diverse use. Basically, this makes their design quite difficult because the different users and user groups have quite different ideas and priorities regarding what they need as information. What the quality label 'fitness for purpose' means in such statistics is therefore anything but trivial. How this problem is addressed is explained in Sect. 2.3.
- In contrast, indicators are closely tied to a specific question and task. In particular, for European policies, it is typical that they provide and promote decision-making

---

[10]See the European statistical programme 2013–2017 (European Union 2011, p. 20).

[11]This consideration and distinction are only recent. It results from the modernisation process of the last 20 years, which will be presented in a later section.

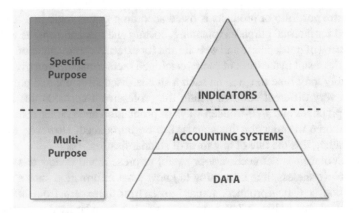

Fig. 2.6 Information pyramid of official statistics. *Source* European statistical programme 2013–2017 (European Union 2011, p. 20)

and governance based on indicators. This has the great advantage that the information requirements are usually very well known. Whether the statistics are 'fit for purpose' can therefore be assessed quite precisely. On the other hand, this closeness to political decisions (sometimes linked to immediate sanctions or other consequences) also carries considerable risks, which are discussed in Chap. 3.

Although this is not a strict and non-overlapping order, the approach helps to better describe the products, the processes and different quality profiles so that they can be better planned, managed and communicated.

The business model of modern statistics includes not only the products but also **statistical services**. This includes, above all, special services for individual users. For example, a statistical office may provide tailored analysis of its data of interest to individual users (e.g. businesses or media) or user groups (industry associations or NGOs). For research and teaching, elaborate work is carried out to allow access to micro-data without jeopardising the confidentiality of individual data.

2.1.5.2 The Product Portfolio

As one would expect from an industry, the products are grouped and managed together in one complete programme, one 'portfolio'. With the help of such a portfolio, internal planning and decisions (priorities) are made possible and a controlling (costs, quality) can be built up. It is very important for communication with users to offer the portfolio in such a form that they can get a good overview of the available information that enables them to make their own choices.

For the sake of clarity, reference is made below to European statistics as an example. In the multiannual planning of the statistical programme, this systematic approach was used to structure the portfolio of products.

The entire portfolio of products is listed according to this logic in a 'catalogue' that is used for internal purposes (planning, costing and management) as well as for the structuring of the database and website and for external communication purposes. Although the result in the form of such a catalogue seems logical, even trivial, it took a remarkably long time to agree on such a standardised structure and presentation. Given the very different cultures within the professional communities in official statistics, practices and well-trodden ways of production and similar resistance had to be overcome in order for this standard to be introduced. However, this is not unusual; rather, it is the fate of any form of standardisation.

At this point, it is not necessary or useful to present and review this catalogue in detail. Nevertheless, it is interesting to know what information such a catalogue contains about statistical products. As can be seen from the excerpt from the European statistics catalogue in Fig. 2.7, the products are given standardised names; they are coupled with the relevant production process, the temporal and regional resolutions are given, and the main users and the legal basis are mentioned.

In the course of modernising official statistics (especially in Europe), it is becoming increasingly important to define modules and services that can be exchanged and shared within an agreed and standardised business architecture. A common product catalogue created, shared and applied by all producers in the statistical system of Europe, therefore, represents a decisive step forward on the path to efficient value chains and close cooperation between the partner institutions. In addition to the product catalogue, a service catalogue will increasingly play a major role.[12]

2.2 Skills and Human Resources

The decisive factor for the quality of statistics is the staff of the statistical institution. First and foremost, of course, this means that the statistical institution must have a sufficient number of sufficiently qualified professionals. In recent decades, there have been major changes in the amount and structure of staff. While the absolute number of employees has tended to decrease, the proportion of academically trained and qualified employees has increased (Figs. 2.8, 2.9 and 2.10).

In this respect, the same development has taken place in official statistics as in other industries, where quantity has been substituted by the quality of the employees. This is the move from Official Statistics 2.0 to 3.0: from a manual to an industrial production of statistical products based on an all-embracing use of information technology.

Another very important consideration is the professional composition of the staff. In the previous sections, the variety of products and processes was explained. Ideally, experts and their knowledge would be available for all products and processes. Of course, that is not possible. In addition, the diversity is too large and the supply on the job market too limited. Above all, however, there is a lack of specialised training, from which graduates could be recruited for official statistics tasks. In addition, it

---

[12] See for example Eurostat (2016b).

16-May-2017   Eurostat Catalogue of Statistical Products 2017   1/11

P.A.	Statistical Products: [Code] Title and Description [1]	Activities of the AWP that produce the Products	Highest frequency [2]	NUTS [3]	Institutional Users	Specific legal base [4]
1	[EU2020] Europe 2020 indicators Europe 2020 headline indicators on employment, research and development, climate change and energy, education, and poverty and social exclusion.	• [EU2020] Europe 2020 indicators	Annual	0	Several DGs	
2	[GovA] Annual government finance • Taxes and tax indicators. • Classification of the Functions of Government (COFOG). • Main aggregates of general government.	• [Gov] Quarterly and Annual Government statistics - Financial and non-Financial Accounts for general government	Annual	0	ECFIN	(EU) 549/2013
2	[GovD&D] Government deficit and debt • Government deficit and debt including annual structure of government debt. • Contingent liabilities and non-performing loans.	• [Gov] Quarterly and Annual Government statistics - Financial and non-Financial Accounts for general government	Annual	0	ECFIN	(EU) 549/2013, Directive 85/2011
2	[GovQ] Quarterly government accounts • Quarterly government non-financial accounts. • Quarterly government financial accounts. • Quarterly Maastricht debt. • Inter-governmental lending (IGL incl. EFSF (European Financial Stability Facility)).	• [Gov] Quarterly and Annual Government statistics - Financial and non-Financial Accounts for general government	Quarterly	0	ECFIN	(EU) 549/2013
2	[MIP] MIP scoreboard Scoreboard of indicators of the Macroeconomic Imbalance Procedure (MIP).	• [MIP] Scoreboard of indicators of the Macroeconomic Imbalance Procedure	Quarterly	0	ECFIN	
3	[EGR] Aggregated statistics based on EGR Aggregated statistics on enterprise groups operating in Europe and their constituent enterprises based on the EuroGroups Register (EGR) micro data. From 2017.	• [EGR] Business registers - coordination and monitoring; EuroGroup Register (EGR) - production and dissemination	Annual	0, Non-NUTS	Several DGs, ECB, ECE/UNECE, UNSD	Reg. 177/2008

Fig. 2.7 Eurostat catalogue of products. *Source* Eurostat (2017b) (extract)

	Number of employees	Lower education (%)	Middle or higher education (%)
1995	2.590	50.2	49.8
2017	1.992	20.4	79.6

Fig. 2.8 Staff in statistical offices—example: Statistics Netherlands (For example, the annual report 2015 of the CBS of the Netherlands expresses this in the following way: "*The increasing complexity and further automation of statistical processes are contributing to the decline in the amount of semi-skilled and unskilled work and the increasing need for more highly educated staff. This is reflected in the composition of the workforce by job grade. In 2015, 77% of staff were in scale 9 or above. By comparison, the proportion in 1995 was only 50%.*" (CBS 2016b, p. 20). Other (unpublished) figures, provided by CBS)

	Number of employees	Clerks (%)	Academics (%)	IT (%)
1995	567	54.2	28.4	17.4
2017	567	11.1	61.2	11.4

Fig. 2.9 Staff in statistical offices—example: statistics Denmark (unpublished figures, provided by Statistics Denmark)

has been shown in recent years that the dynamics of change are so great that, in any case, vocational training and internal training on the job are the more important qualification methods.

Which qualifications and professional orientations are actually needed? Which competences should be available among the team in the statistical office?

First, in the classic field of statistical production (including the development of methods, products and processes), one could falsely assume that primarily statisticians would come into play, who are familiar with survey techniques. This is true, but only in the areas in which data about surveys are originally collected, e.g., in a wide range of social statistics or business statistics.

With the increasing importance of administrative data and registers as a source of official statistics, the profile of requirements has already changed in recent years. Of course, if work processes no longer start with collecting data, and if instead existing data needs to be analysed, filtered and aggregated to meaningful information, the job profile will change accordingly. With the omnipresence of Big Data the working conditions will undergo even further dramatic change, which again requires new personal skills, knowledge and experience.

Fig. 2.10 Staff in statistical offices—example: statistics Denmark (unpublished figures, provided by Statistics Denmark)

This does not mean that training in survey methodology is or will become irrelevant. Rather, paths must be taken in which these competencies are embedded in methods of data sciences and the management of complex production processes.

Second, it requires expertise in the area of accounts, be it the macro-economic, social (health, education …) or environmental accounts. A solid education in economics with an emphasis on the empirical focus on National Accounts is absolutely necessary here.

Third, and this is still a relatively recent domain, knowledge in the field of indicator methodology is needed. Needless to say, this requires a mix of statistical methods, communication skills and sensitivity for the policy dimension of the specific indicator (being aware of the respective opportunities and threats related to the closeness to policy-making).

Fourth, statistical office staff is expected to have sufficient knowledge of the area of application for which they are responsible. This can be a more specific and narrow area, such as agricultural, energy or health, or a wider one, such as business cycle, labour market or Sustainable Development.

Fifth, it is of course of strategic importance to have the necessary expertise in the fields of information and communication technologies available. However, due to the very dynamic development, it is increasingly difficult or even impossible to maintain this expertise in-house. An outsourcing of such services and the corresponding personnel capacity is essential.

Sixth, today it is more important than ever to have specialists in the field of communication and media in the team.

Finally, in today's administrations, it is not only the classical administrators but also skilled professionals of modern management (quality management, controlling, cost accounting, etc.) who play an important role.

However, the actual composition of the staff depends very much on external factors and framework conditions: Are appropriate training courses offered at the universities? Is the statistical office attractive and competitive on the (local) labour market? In order to improve conditions in this regard, European statistics has launched 'EMOS',[13] a Master's degree programme designed to better prepare graduates for their employment in official statistics.

The difficulties that official statistics face in human resources are changing over time. For example, it has become increasingly problematic to find well-trained economists for working in National Accounts; apparently, the empirical dimension plays only a sub-ordinate role in today's economics studies. In the context of digitisation, developments will take place, which must also be reflected in the composition of personal and professional skills. Finally, in the future, more attention will have to be paid to the interplay and interactions of statistics and society, which also requires corresponding specialist capacities. This corresponds with the topic of the present work.

---

[13] https://ec.europa.eu/eurostat/cros/content/what-emos_en.

For the sections that follow (in particular, the section on quality), it is important to understand the interaction between the composition of the staff and the statistical culture that is emerging in different areas of work. As a rule, two communities are represented and these could hardly be more different in their views and ways of working.

On the one hand, there are the survey statisticians (especially in the field of social statistics), whose quality reliance is based on the fact that the entire production process of survey design (from data collection and processing to the generation of aggregated results) is under their control. Here, quality aspects such as reliability and punctuality are in the foreground, while a complete coverage of a topic or consistency is seen as of minor importance. Such an approach is called 'micro'.

On the other hand, in the field of accounting, the primary concern is a complete and consistent picture of a situation or a subject matter area, while accuracy in details plays a minor role. Such approaches are called 'macro'.

From these two approaches and cultures arise, in some cases, considerable (micro-macro) differences in the statistical results on the same topic.[14] Because of this, it is difficult to subsume the quality of statistical products under a single definition.

2.3 Quality in Official Statistics

2.3.1 Quality—An Old Objective—A Young Concept

In order for official statistics to function as a language, a 'boundary object' (Stamhuis 2008; Saetnan et al. 2011) for all kinds of societal interactions and decision-making, it is essential that the quality of statistical products and services is outstanding, an authority in itself. For Porter, '*the language of quantification may be even more important than English in the European campaign to create a unified business and administrative environment*' (Porter 1995, p. 77). This is the brand-mark and the competitive advantage of official statistics. Once this authority is undermined, be it through real quality problems or only through perception, trust in official statistics will be replaced by suspicion and statistics will become part of political fights and games. Against this background, it is important to define quality of statistics with a much wider scope, including not only the production but also the use side of statistical information and how these two sides are interacting in a dynamic relationship.

As a consequence, the approach to quality in official statistics has changed radically over the past two decades. According to today's prevailing opinion, statistics must be suitable for a particular utilisation: this is the criterion 'Fitness for Purpose!' However, unlike the earlier producer-related view and definition of quality, this new objective leads to a very complex world in which simple and one-dimensional solutions are no longer possible or appropriate.

---

[14]See for example the case of income, consumption and wealth (Brandolini 2016).

Central to the remarks is to understand statistics as products, products of a larger whole (the portfolio), produced under given conditions and constraints and aimed at serving a not necessarily sharply defined group of users. For every single product as well as for the entire portfolio, it is important to find a ('Pareto') optimal solution, meaning to achieve the best of all possible solutions for each statistic and for the statistical program as a whole (Radermacher 1992, 1999a). This may sound abstract and difficult. However, it becomes plausible and practically solvable in an evolutionary process, with year-by-year changes in planning and production.

In European statistics, the first systematic steps in the area of statistical quality were made at the end of the 1990s through cooperation in the ESS Leadership Group (LEG); on quality initially, the LEG was struggling with difficulties inherent in the convergence of two schools of thought: classical approaches from statistical methodology and approaches from industrial quality management. It was very much in the spirit of W. E. Deming's[15] view on 'profound knowledge', quality management and learning organisations, which the LEG had finally elaborated in a synthesis report, including 21 recommendations for European statistics (Lyberg et al. 2001).

2.3.2   Quality Objectives and Means to Reach Them

Unfortunately, there is no unified glossary of quality terms in official statistics. A search on the corresponding page of the OECD gives a total of 131 hits.[16]

Therefore, a more general approach to the topic of quality in statistics will be made here, before going into the various aspects in detail.

Statistical information should, as far as possible, meet three different requirements (see Fig. 2.11).

First, it should provide information about a phenomenon that is relevant to answering current questions. Statistics that interest nobody cannot be part of the tight budget of a statistical office. Here, of course, one is immediately confronted with the crucial problem that concerns the selection of topics and issues to which this relevance relates. So, who determines the statistical programme in the end?

Second, the statistics should be supported by a theory, i.e. they should meet scientific standards. In this regard, it has to be clarified which theory is meant here. For the National Accounts, the case is comparatively clear: they are closely linked to the macro-economy. Less clear, however, is this objective in the remaining areas, even if close links exist between empirical research and official statistics. In the more recent areas (environment, Sustainable Development), the situation is particularly difficult because of the various disciplines involved.[17]

Third, of course, statistics should meet the criteria of measurability, they should be reliable, punctual, comparable and accessible, to name but a few.

---

[15] See Deming (2000).

[16] See OECD https://stats.oecd.org/glossary/ (Quality).

[17] For the complexity inherent to multidisciplinarity see (Klein 2004).

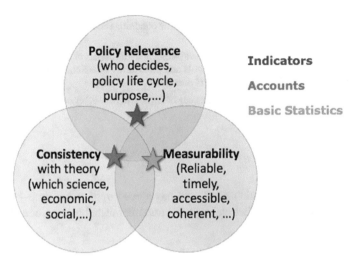

Fig. 2.11 Dimensions of statistical quality

Apparently, different aspects and categories play a role in the (multidimensional) definition of statistical quality. Additionally, because it is not possible, under normally limited circumstances (costs, time, staffing, willingness to provide information), to fully meet all these objectives at the same time, priorities must be set, e.g., in the statistics programme or in the selection of statistic variables.

In this respect, it is advantageous that in the portfolio of statistics different products are included, which embody with their special quality profile, in different ways, the three objectives of relevance, consistency and measurability. While basic statistics in their great diversity are very much aligned with the goal of measurability, National Accounts focus on scientific consistency within a specific theory. In turn, indicators are closer to the goal of relevance. In this respect, the different types of statistical products are not only located at different levels of the information pyramid, but also complement each other.

In the European statistics the basic elements of such a quality approach are manifested in Regulation 223 (European Union 2015: Art 12):

European statistics shall be developed, produced and disseminated on the basis of uniform standards and of harmonised methods. In this respect, the following quality criteria shall apply:

(a) 'relevance', which refers to the degree to which statistics meet current and potential needs of the users;

(b) 'accuracy', which refers to the closeness of estimates to the unknown true values;

(c) 'timeliness', which refers to the period between the availability of the information and the event or phenomenon it describes;

(d) 'punctuality', which refers to the delay between the date of the release of the data and the target date (the date by which the data should have been delivered);

(e) 'accessibility' and 'clarity', which refer to the conditions and modalities by which users can obtain, use and interpret data;

(f) 'comparability', which refers to the measurement of the impact of differences in applied statistical concepts, measurement tools and procedures where statistics are compared between geographical areas, sectoral domains or over time;

(g) 'coherence', which refers to the adequacy of the data to be reliably combined in different ways and for various uses.

2.3.3   Code of Practice

In European statistics, in this regard, one has agreed upon a structure—one could also say classification—further elaborated[18] and manifested in the Code of Practice (Eurostat 2011):

> The European Statistics Code of Practice sets out 15 key principles for the production and dissemination of European official statistics and the institutional environment under which national and Community statistical authorities operate. A set of indicators of good practice for each of the 15 principles provides a reference for reviewing the implementation of the Code.

The European Statistics Code of Practice was adopted by the Statistical Programme Committee in 2005 and was revised by the European Statistical System Committee in September 2011 and 2017 (Eurostat 2018b) (Fig. 2.12).

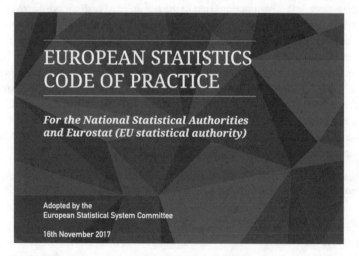

Fig. 2.12 European Statistics Code of Practice

---

[18]EU Regulation 223: *"The statistical principles set out in this paragraph are further elaborated in the Code of Practice"* (European Union 2015: Art 12).

ES Code of Practice:

- Institutional environment
  - Institutional and organisational factors have a significant influence on the effectiveness and creditability of a statistical authority developing, producing and disseminating European Statistics. The relevant issues are professional independence, mandate for data collection, adequacy of resources, quality commitment, statistical confidentiality, impartiality and objectivity.
- Statistical processes
  - European and other international standards, guidelines and good practices are fully observed in the processes used by the statistical authorities to organise, collect, process and disseminate European Statistics. The credibility of the statistics is enhanced by a reputation for good management and efficiency. The relevant aspects are sound methodology, appropriate statistical procedures, non-excessive burden on respondents and cost effectiveness.
- Statistical output
  - Available statistics meet users' needs. Statistics comply with the European quality standards and serve the needs of European institutions, governments, research institutions, business concerns and the public generally. The important issues concern the extent to which the statistics are relevant, accurate and reliable, timely, coherent, comparable across regions and countries, and readily accessible by users.

This short summary makes it clear that the quality concept follows the three dimensions that were already introduced at the beginning: Who? (Institutions), How? (Processes) and What? (Products). Like the Code of Practice itself, the way it is implemented is significantly inspired by the methods of Total Quality Management (TQM). This is mainly reflected in the Quality Assurance Framework (QAF) (Eurostat 2018d).

Box 2.1 Quality criteria for European statistics[19]

Institutional environment	Statistical processes	Statistical output
1. Professional independence 1bis. Coordination and cooperation 2. Mandate for data collection 3. Adequacy of resources 4. Commitment to quality 5. Statistical confidentiality 6. Impartiality and objectivity	7. Sound methodology 8. Appropriate statistical procedures 9. Non-excessive burden on respondents 10. Cost effectiveness	11. Relevance 12. Accuracy and reliability 13. Timeliness and punctuality 14. Coherence and comparability 15. Accessibility and clarity

It is essential to the success of quality management that the governance (the 'who?') is thoroughly thought through and implemented in a timely manner. With the Code of Practice, e.g., in Europe, the committees of statistics are being reformed.

---

[19]The 2017 edition of the CoP is based on 16 principles.

Among other things, a supervisory board (European Governance Advisory Board[20]) was established. In addition, peer reviews[21] for the producers of European statistics were carried out at longer intervals.

For the part of European statistics produced by the ECB and the national central banks, a similar quality framework is in place (ECB 2018).

The existence of codes of conduct in official statistics is still comparatively new, introduced in the past three decades. First, ethical standards for the professional statistician were adopted by the International Statistical Institute (ISI) in 1985 (ISI 2018); in 2010, the declaration of professional ethics followed (ISI 2010).

Box 2.2 ISI Professional Ethics[22]
"Our shared professional values are respect, professionalism, truthfulness and integrity."

Ethical Principles

- Statisticians should pursue **objectivity** without fear or favor, only selecting and using methods designed to produce the most accurate results. …
- **Clarifying Obligations and Roles**: … statisticians should take care to stay within their area of competence, and seek advice, as appropriate, from others with the relevant expertise.
- **Assessing Alternatives Impartially:** Available methods and procedures should be considered and an impartial assessment provided to the employer, client, or funder of the respective merits and limitations of alternatives, along with the proposed method.
- **Conflicting Interests**: Statisticians avoid assignments where they have a financial or personal conflict of interest in the outcome of the work. …
- Avoiding **Preempted Outcomes**: Any attempt to establish a predetermined outcome from a proposed statistical inquiry should be rejected, …
- Guarding **Privileged Information**: Privileged information is to be kept confidential. This prohibition is not to be extended to statistical methods and procedures utilized to conduct the inquiry or produce published data.
- Exhibiting **Professional Competence**: Statisticians shall seek to upgrade their professional knowledge and skills, …
- Maintaining **Confidence in Statistics**: In order to promote and preserve the confidence of the public, statisticians should ensure that they accurately and correctly describe their results, including the explanatory power of their data. …
- **Exposing and Reviewing Methods and Findings**: Adequate information should be provided to the public to permit the methods, procedures, techniques, and findings to be assessed independently.

---

[20]See ESGAB (2018).

[21]See Eurostat (2018c).

[22]ISI (2010).

- **Communicating Ethical Principles**: In collaborating with colleagues and others in the same or other disciplines, it is necessary and important to ensure that the ethical principles of all participants are clear, understood, respected, and reflected in the undertaking.
- Bearing **Responsibility for the Integrity of the Discipline**: Statisticians are subject to the general moral rules of scientific and scholarly conduct: they should not deceive or knowingly misrepresent or attempt to prevent reporting of misconduct or obstruct the scientific/scholarly research of others.
- Protecting the **Interests of Subjects**: Statisticians are obligated to protect subjects, individually and collectively, insofar as possible, against potentially harmful effects of participating. ....

The need for a set of principles governing official statistics became apparent at the end of the 1980s when countries in Central Europe began to change from centrally planned economies to market-oriented democracies. It was essential to ensure that national statistical systems in such countries would be able to produce appropriate and reliable data that adhered to certain professional and scientific standards. Towards this end, the Conference of European Statisticians developed and adopted the Fundamental Principles of Official Statistics in 1991 ... a milestone in the history of international statistics was reached when the United Nations Statistical Commission at its Special Session of 11–15 April 1994 adopted the very same set of principles – with a revised preamble – as the United Nations Fundamental Principles of Official Statistics. (UNSD 2018)

The current version of the UN Fundamental Principles (see Box 2.3) was endorsed by the UN General Assembly in its resolution 68/261 of 29 January 2014 (United Nations 2014).

Box 2.3 UN Fundamental Principles of Official Statistics[23]

- **Principle 1**. Official statistics provide an indispensable element in the information system of a democratic society, serving the Government, the economy and the public with data about the economic, demographic, social and environmental situation. To this end, official statistics that meet the test of practical utility are to be compiled and made available on an impartial basis by official statistical agencies to honour citizens' entitlement to public information.
- **Principle 2**. To retain trust in official statistics, the statistical agencies need to decide according to strictly professional considerations, including scientific principles and professional ethics, on the methods and procedures for the collection, processing, storage and presentation of statistical data.
- **Principle 3**. To facilitate a correct interpretation of the data, the statistical agencies are to present information according to scientific standards on the sources, methods and procedures of the statistics.
- **Principle 4**. The statistical agencies are entitled to comment on erroneous interpretation and misuse of statistics.

---

[23] Source United Nations (2014).

- **Principle 5**. Data for statistical purposes may be drawn from all types of sources, be they statistical surveys or administrative records. Statistical agencies are to choose the source with regard to quality, timeliness, costs and the burden on respondents.

- **Principle 6**. Individual data collected by statistical agencies for statistical compilation, whether they refer to natural or legal persons, are to be strictly confidential and used exclusively for statistical purposes.

- **Principle 7**. The laws, regulations and measures under which the statistical systems operate are to be made public.

- **Principle 8**. Coordination among statistical agencies within countries is essential to achieve consistency and efficiency in the statistical system.

- **Principle 9**. The use by statistical agencies in each country of international concepts, classifications and methods promotes the consistency and efficiency of statistical systems at all official levels.

- **Principle 10**. Bilateral and multilateral cooperation in statistics contributes to the improvement of systems of official statistics in all countries.

2.3.4 Quality Management, Quality Assurance

In combination with European statistics quality standards and practices,[24] individual statistical offices use methods and frameworks of modern quality management (TQM), which are also customary in industry, such as EFQM[25] or Lean Six Sigma.[26] Total Quality Management, in its various conceptual variants, always follows a holistic approach that captures a company in all its facets, from goal setting and strategy to input and processes to results, outputs and results of various forms.

TQM models have proven themselves in practice and guarantee a systematic approach that incorporates all factors of a continuous improvement process. As an example, the EFQM model makes it clear with its criteria[27] that TQM is not purely about optimising and controlling production processes. Rather, TQM is at the forefront of the entire company. The basic ideas of a comprehensive and systemic quality management are based on thinkers like Russell L. Ackoff, Peter Drucker and above all W. Edwards Deming. Deming, who was a statistician, by the way, emphasised the importance of managers not interpreting their role purely technically or economically. If they want to be successful, i.e. to produce excellent quality, they must understand their company, its employees, the interrelationships and backgrounds and much more in a profound knowledge (see Chap. 3). When producing statistics, it is therefore also important to have in-depth professional knowledge of the managers

---

[24]See for example ISTAT (2018) or INE Portugal (INE 2018).

[25]See for example Destatis (2018) or Eurostat (2018d).

[26]See for example Statistics Netherlands (CBS 2016a).

[27]See EFQM (2013, p. 9).

and quality of management. Management of quality means first of all quality of management.

Another important branch in the field of statistics quality has to do with cooperation and above all international comparability. In Europe, important policy decisions and significant financial flows are immediately influenced by and directly dependent on the comparability and robustness of national statistics. These include, in particular, Member States' contributions to the EU budget and the Maastricht Treaty criteria and indicators for monitoring the (excessive) deficit of the Member States. In these political areas, quality assurance frameworks have been deeply embedded in the European legal system: the statistical methods have been agreed and fixed in detail, reporting obligations have been standardised, and audit-like supranational controls have been introduced. In addition to the statistics with an average quality profile, a category of indicators has emerged as a result of the evidence-based decision-making, which has a special authority and very special significance for politics and public discourse. It can be predicted without great uncertainty that this trend will continue and with it the political expectations vis-à-vis this premium class of statistics. The extent to which the politicians' willingness to pay will thus increase is just as uncertain, as is the answer to the question of whether this involves risky assignments to statistics and unrealistic expectations of quantifiability.

International statistics are increasingly being used in the service of the mainstream logic of 'evidence-based decision-making'. If statistics are given more responsibility, it will result in a chain of consequences with quality assurance measures[28] and, unfortunately, also in attempts to circumvent them.[29]

The risks arising from the closer relationship between indicators and policy decisions have been addressed by a package of measures[30] that go beyond traditional quality reporting. These include, above all, so-called Peer Reviews,[31] but also, in selected areas of indicators of particular political significance, surveillance and control measures and, in the case of public finance statistics, even possible sanctions in the event of non-compliance with statistical standards.

In summary, it can be said that the development of the topic of quality in statistics since the early 1990s mirrors the evolution of statistics over this entire period. It is above all the improvements in productivity through modernisation and industrialisation taking place throughout this era, but also the societal and political framework conditions that have had a lasting effect on and changed the statisticians' relationship to 'quality'.

---

[28] See for example IMF (2017, 2018) or OECD (2015).

[29] See for example Eurostat on Greece (European Commission 2010) or IMF on Argentina (IMF 2016).

[30] See in particular "Towards robust quality management for European Statistics" (European Commission 2011).

[31] See Eurostat (2018c).

2.3.5 *Evolution and Continuous Adaptation*

2.3.5.1 The Learning Cycle, Continuous Improvement

The great diversity and complexity of the subject 'quality' in official statistics have become clear in the preceding sections. Simple and quick solutions to these questions are therefore not expected and nor would they be appropriate. Rather, it becomes very clear that the wide range of processes and products in the statistical factory can only develop over longer periods of time, and that this feature of continuous improvement will continue to be inseparable from official statistics in future.

It is therefore important to make a virtue of this fact by finding a dynamic development and balance between preserving and innovating in terms of W. E. Deming's quality management (see Fig. 2.13):

> The PDSA Cycle (Plan-Do-Study-Act) is a systematic process for gaining valuable learning and knowledge for the continual improvement of a product, process, or service. … The cycle begins with the Plan step. This involves identifying a goal or purpose, formulating a theory, defining success metrics and putting a plan into action. These activities are followed by the Do step, in which the components of the plan are implemented, such as making a product. Next comes the Study step, where outcomes are monitored to test the validity of the plan for signs of progress and success, or problems and areas for improvement. The Act step closes the cycle, integrating the learning generated by the entire process, which can be used to adjust the goal, change methods, reformulate a theory altogether, or broaden the learning – improvement cycle from a small-scale experiment to a larger implementation Plan. These four steps can be repeated over and over as part of a never-ending cycle of continual learning and improvement. (Deming Institute 2018)

Fig. 2.13 PDSA cycle

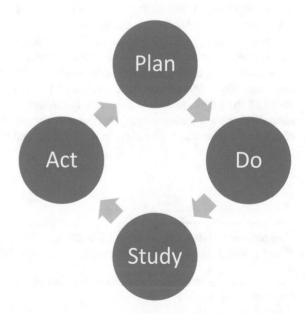

In the spirit of this continuous and systematic process of learning and improving official statistics, dialogue with users and interaction with science is crucial. The planning of official statistics at all levels (variable, product and programme) has to be understood and organised as an evolutionary process, as a sequence of learning cycles and feedback loops.

Over time, changes might be started from all three angles. New demands and political issues trigger new statistical developments, as new data sources or new methodologies do. Historically, it can be observed that these driving forces are also mutually influencing each other, thus stimulating new episodes in official statistics (Desrosières 1998).

It is therefore essential to link the communication process of today with the development of statistics for tomorrow. Partly, this loop could be a short one, if user feedbacks can lead to quick fixes and improvements in services. Partly, however, it might take time, since changes in a programme need profound preparations and even more profound developments (Fig. 2.14).

This evolutionary development of statistics is confronted with several limiting factors, which could be practical limitations, such as:

- Clandestine, non-observable phenomena
- Statistical items in future and elsewhere, relevant for decisions now and here (e.g., capital goods, depreciation, trade chains, Sustainable Development)
- Values and prices for non-market-goods (can we simulate non-existing markets?)
- Limitations by resource or time constraints; in cases, where only a limited amount of information and data is available or where limited time is given for the decision-making process
- Limitations could also relate to the understanding and use of data and information
- Innumeracy, statistical and data illiteracy

Fig. 2.14 Statistical learning process

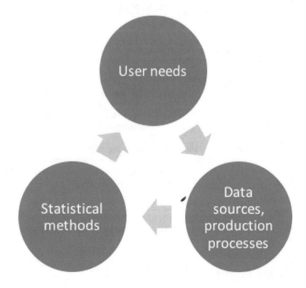

- High and too high expectations
- (No) Appetite for high-quality information (Davies 2017).

Since the beginning of official statistics in the nineteenth century, the boundaries have been substantially stretched. Continuous improvement has opened new opportunities so that today many subjects (e.g., quality of life) that were impossible to observe only a few years ago are fully integrated elements of the standard statistical programme. Nevertheless, it is crucial to understand that basic principles must be respected, if the fundament on which trust in official statistics is built is not to be damaged. This is, for example, the reason to refrain from monetising natural resources and their services, if they are not valued by market transactions.

2.3.5.2 Planning and Decision Procedures, Consultation of Users, Governance

Official statistics is a special application and form of statistics that belong to the public infrastructure of (modern) states. Working methods in official statistics reflect both their political and administrative position as well as the status and development of societies (i.e. the specific relationship between state and citizens).

How is public infrastructure planned and decided? This is generally an important issue, as it is about providing the greatest possible value and return on taxpayers' money. The way the statistics programme is decided reflects what is called governance (and what Foucault would have called *gouvernementalité*[32]) and the status of the relationship between the state and its institutions on the one side and civil society (the citizens, interest groups) on the other.

Public statistics in a modern and democratic state has to be benchmarked against principles of good governance in the public sector, out of which the overarching and most important ones are (Fig. 2.15).

Behaving with integrity, demonstrating strong commitment to ethical values, and respecting the rule of law	Public sector entities are accountable not only for how much they spend, but also for how they use the resources under their stewardship. This includes accountability for outputs, both positive and negative, and for the outcomes they have achieved. Public sector entities are accountable to legislative bodies for the exercise of legitimate authority in society. This makes it essential that each entity as a whole can demonstrate the appropriateness of all of its actions and has mechanisms in place to encourage and enforce adherence to ethical values and to respect the rule of law.
Ensuring openness and comprehensive stakeholder engagement	As public sector entities are established and run for the public good, their governing bodies should ensure openness in their activities. Clear, trusted channels of communication and consultation should be used to engage effectively with all groups of stakeholders, such as individual citizens and service users, as well as institutional stakeholders.

Fig. 2.15 Overarching principles of good governance. Adapted from IFAC (2014)

---

[32]See for example Foucault (1991).

At least for European statistics, it can be stated that the first criterion is already taken very seriously. Both the statistical programmes and the single statistical acts go through cumbersome legislative procedures, which place extremely high demands on ex-ante impact assessment and consensus-building prior to final decision-making. Of course, such ambitious legal procedures also have a price, namely the lack of adaptability and dynamics of the statistical system. It should also be mentioned that important control mechanisms (quality reports to the European Parliament, review by the European Court of Justice, etc.) have been set up and are in place for the implementation of European statistics.

For the second criterion (stakeholder engagement), however, the status quo leaves some room for improvement: In a modern and democratic definition, official statistics is no longer a knowledge tool in the hands of the powerful and mighty. Rather, it must follow principles of neutrality and impartiality, whereby this information infrastructure becomes an important democratic pillar, equally available and accessible for everyone.

The interaction with stakeholders must be governed by principles of transparency, democratic control/supervision and public/legislative conventions. In particular, the programme of work must emerge from a democratic decision-making process, at the end of which a choice is made in favour of the 'Pareto-optimal' composition of statistical tasks. Priority setting in this context has an important role to play, as it must facilitate the annual adaptation of the programme following changes in user needs.

The way in which this consultation and decision-making have been organised so far relies mainly on the functioning of 'official' procedures concerning the preparation of legislation and political decisions. Modern societies, however, ask for more— more in terms of wider consultation (more room for all active contributions from civil societies), new forms (collection of user needs through social media) and speed (quicker adaptation of the programme).

2.4 National, International and European Statistics

As explained at the beginning of this chapter, official statistics and its history are closely linked to the nation state. This particular form and institution of governance emerged in the eighteenth and nineteenth century. Since this time, there has been a strong need for statistics in all forms and for all kind of political decisions.

In order to be able to describe official statistics more precisely, it is therefore necessary to understand the methodical orientation towards the information needs and the institutional integration into the administrative system of the nation state. The entire structure, methodology, design of surveys, the conception of macro-economic accounts, indeed everything in today's official statistics, bears a 'national stamp'.

National Accounts provide an overarching ordering system within official statistics. The latest versions of the System of National Accounts SNA 2008 and the European System of Accounts ESA 2010 are the accumulated result of more than six decades evolution of statistical knowledge, emerging from intensive collaboration between specialists in accounting and statistical methodology, scientists and users. These are international standards of enormous importance for all official statistics. Hundreds, even thousands of difficult considerations, methodical decisions and conventions have melted and flowed into them. The most recent version of the *ESA Handbook* contains 650 pages of definitions and guidelines of all kinds (Eurostat 2013b).

All of these conventions and standards are geared towards and optimised for one goal, namely to quantify the economic activity (production, income, consumption) of a state in a temporal period (year, quarter) as comprehensively and precisely as possible. A closer look at some of the National Accounts aggregates[33] explains this:

- **Gross domestic product at market prices** (GDP): Gross domestic product at market prices is the final result of the production activity of resident producer units. It can be defined in three ways:

- production approach: GDP is the sum of gross value added of the various institutional sectors or the various industries plus taxes and less subsidies on products (which are not allocated to sectors and industries). It is also the balancing item in the total economy production account;
- expenditure approach: GDP is the sum of final uses of goods and services by resident institutional units (final consumption and gross capital formation), plus exports and minus imports of goods and services;
- income approach: GDP is the sum of uses in the total economy generation of income account (compensation of employees, taxes on pro- duction and imports less subsidies, gross operating surplus and mixed income of the total economy).
- By deducting consumption of fixed capital from GDP, we obtain **net domestic product at market prices** (NDP).
- **National income** (at market prices): Gross (or net) national income (at market prices) represents total primary income receivable by resident institutional units. It equals GDP minus primary income payable by resident institutional units to non-resident institutional units plus primary income receivable by resident institutional units from the rest of the world.
- **Current external balance**: The balancing item in the external account of primary income and current transfers represents the surplus (if it is negative) or the deficit (if it is positive) of the total economy on its current transactions (trade in goods and services, primary incomes, current transfers) with the rest of the world.

Following these definitions, it becomes obvious that the core of National Accounts is anything but trivial. The objective of quantifying the economy of one country in one specific year can only be achieved by establishing in principle what is to be regarded as an activity, which activity belongs to that country and under what conditions it is attributed to that specific year. This, in turn, requires ancillary calculations that allow the demarcation between that country and the rest of the world (e.g., import/export, balance of payments), and which allows for offsetting the periodic allocations (e.g., depreciation).

---

[33]ESA 2010 (Eurostat 2013b, pp. 273–74).

This brief introduction to the concepts of National Accounts should suffice to illustrate a paradigm that is fundamental to the structure and results of official statistics. The first and most important task of official statistics is to satisfy information needs at national level. A comprehensive definition of official statistics must therefore take this paradigm into account. The three dimensions of national official statistics are:

- Temporal dimension: a fixed time period (very often a year, but sometimes also a quarter or a month) or a fixed date (for example, for the population census)
- Spatial dimension: a country (political delimitation) or a region (province, local unit according to administrative delimitation, i.e. NUTS the 'Nomenclature of Territorial Units for Statistics')
- Measurement object: resident population and their activities; methodologies (variables, classifications, sampling schemes, etc.) designed to address national needs and priorities (Fig. 2.16).

What is strong at the national level, however, is a difficulty in terms of comparability between the statistics of different countries. Because this comparability has grown in importance over the past decades, statistical methods have been standardised, especially nomenclatures and guidelines for general survey methods on important statistics such as the census. Nonetheless, the national shaping of official statistics limits international comparability, in particular, for all statistics that are produced on the basis of administrative sources, thus strongly depending on the respective national legal bases. For instance, labour market statistics, as prepared by the National Employment Agencies, are geared to the national labour market policy. This greatly restricts their use for international comparisons. In Europe, the need for comparability across EU Member States is particularly high, because the statistical indicators are incorporated into European policies and are partly associated with considerable consequences. For this reason, in important areas (such as the labour market), standardised sample surveys have been introduced that consider international definitions; for example, the Labour Force Survey (LFS).[34] Another

Fig. 2.16 National statistics

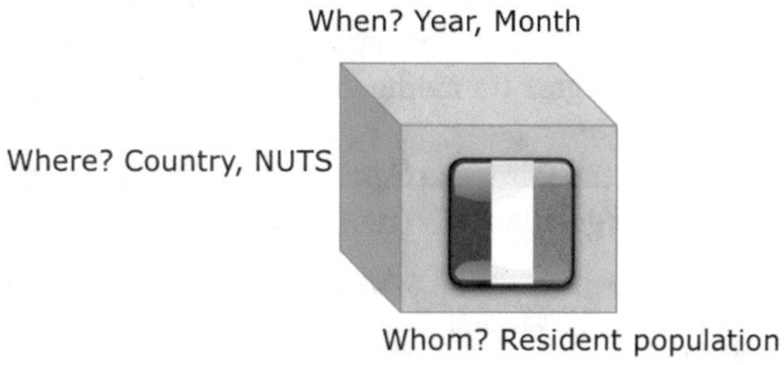

When? Year, Month

Where? Country, NUTS

Whom? Resident population

---

[34] See definition here http://ec.europa.eu/eurostat/statistics-explained/index.php/Glossary:Labour_force_survey_(LFS).

type of solution to this problem is represented by the Harmonised Index of Consumer Prices (HICP),[35] which is derived from the National Consumer Price Indices using a standardised methodology.

These two approaches, namely standardisation and (European) sample surveys, have contributed to make great progress for European statistics in recent years, by methodologically unifying national statistics, thus making their results much more comparable.

However, this progress in terms of international comparability has a price in other categories of quality. From the point of view of users interested in national results only, the international compromise appears to be partly unsatisfactory because it does not meet their information needs as well as they were used to from purely national statistics. In cases where (as in the labour market) national and international statistics are produced side by side, there is regular user confusion, even conflicts in cases where the results allow different interpretations.

Above all, these few examples make one point clear: statistics optimised for national needs have their limits for international issues or comparisons, and vice versa, internationally agreed (or even produced) statistics do not necessarily meet all national requirements. After all, statistical quality is not absolute and not dependent solely on production. Rather, the quality concept is based on the fitness for purpose. If national needs are prioritised, as has been the case in the past, this has consequences at an international level. The transition to more internationally oriented statistical systems is necessary in the face of globalisation, but it will be iterative and will not be without difficult adjustment phase.

Official statistics produced by international organisations (most notably UN and OECD) are based essentially on the approach of coordinating and standardising national statistics. Even in European statistics, which by the very existence of the EU has a different (supranational) claim, the basic architecture of the European Statistical System does not look any different; the 'system' is a network of autonomous national partners.

This means that statistical design strives for a compromise between national and European data requirements, while the data collection and the computation of statistics remains within national competence. In the vast majority of cases, it has been sufficient to agree on methodological standards that are adhered to by all partners (Fig. 2.17).

In 2009, the European Statistical System (ESS) was established by the European Statistics Regulation 223 (European Union 2015: Art 4):

> The European Statistical System (ESS) is the partnership between the Community statistical authority, which is the Commission (Eurostat), and the national statistical institutes (NSIs) and other national authorities responsible in each Member State for the development, production and dissemination of European statistics.

> Member States collect data and compile statistics for national and EU purposes. The ESS functions as a network in which Eurostat's role is to lead the way in the harmonisation of

---

[35]See   definition   here   http://ec.europa.eu/eurostat/statistics-explained/index.php/Glossary: Harmonised_index_of_consumer_prices_(HICP).

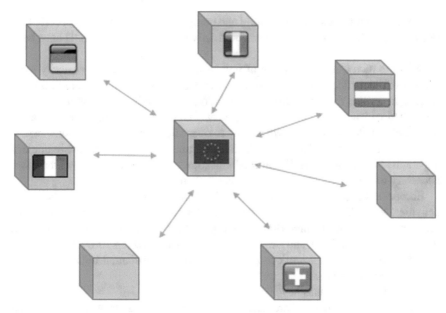

Fig. 2.17 European Statistical System

statistics in close cooperation with the national statistical authorities. ESS work concentrates mainly on EU policy areas – but, with the extension of EU policies, harmonisation has been extended to nearly all statistical fields. The ESS also coordinates its work with candidate countries and at European level with other Commission services, agencies and the ECB and international organisations such as OECD, the UN, the International Monetary Fund and the World Bank.[36]

The concept of a 'partnership', which was introduced already in the nineties in projects launched by the ESS, became in 2009 a key feature of the ESS itself and embodies a compromise which, on the one hand, introduces a 'system' that unites representatives of the European Commission (Eurostat) and the Member States (National Statistical Authorities) in one institutional framework. On the other hand, this system is not organised very stringently or even hierarchically, but as a decentralised network of equal partners.

European political developments have further asked for customised statistical solutions. The introduction of a European single market has created the need for another form of external trade statistics (i.e. Intrastat[37]), the Maastricht Treaty asked for special statistical monitoring of the excessive deficit (i.e. EDP statistics[38]), the

---

[36]Definition: http://ec.europa.eu/eurostat/web/european-statistical-system.

[37]Definition: http://ec.europa.eu/eurostat/statistics-explained/index.php/Glossary:Intrastat.

[38]Definition: http://ec.europa.eu/eurostat/statistics-explained/index.php/Glossary:Excessive_deficit_procedure_(EDP).

European Central Bank requested solid and comparable price statistics (i.e. HICP[39]) (Radermacher 2012), European Household Surveys (EU-SILC[40], LFS[41]) have been established.

With the increasing importance of policy-making and hardening austerity measures, such statistical obligations of Member States have become more and more legally prescribed through European regulations or directives. First of all, this was associated with advantages for all actors. However, the body of European legislation in the field of statistics has grown rapidly to over 300 individual regulations, resulting in a completely inadequate, inflexible, fragmented and incoherent regulatory apparatus. Parallel to the technical modernisation of the processes, it was therefore necessary to simplify and standardise this set of rules.

European statisticians were at the forefront of the international modernisation activities. With 'Vision 404' (Eurostat 2009), a strategy for the next years was outlined that contained all three dimensions: process, product and governance. With the communication 'GDP and Beyond' (European Commission 2009) the European Commission has set up a work programme aiming at substantial improvements of statistical information.

Meanwhile, these plans have been implemented:

The governance of the European Statistical System was substantially revised (Eurostat 2015; Radermacher 2014). The multiannual programme was adapted to the new business model (European Union 2013). The accounting layer (European System of Accounts 2010) has been modernised and broadened (Eurostat 2013c, 2016a). Basic statistics have been re-engineered (e.g., demographic statistics, population census, consumer price statistics, integrated social, business and agricultural statistics). Close cooperation amongst the partners in the European Statistical System was the enabling factor to forcefully implement the strategy that was outlined in the ESS Vision 2020 (Eurostat 2013a).

The governance framework set by the European Statistical System has been successfully used and filled by the modernisation measures of recent years. However, there are increasing difficulties that point to the substantial and principal shortcomings of the national statistics paradigm, which cannot be fixed with the toolkit used so far. Chapter 4 will elaborate on these shortcomings in more detail and explain the approaches to overcome them.

---

[39]Definition: http://ec.europa.eu/eurostat/statistics-explained/index.php/Glossary:Harmonised_index_of_consumer_prices_(HICP).

[40]Definition: http://ec.europa.eu/eurostat/statistics-explained/index.php/Glossary:EU_statistics_on_income_and_living_conditions_(EU-SILC).

[41]Definition: http://ec.europa.eu/eurostat/statistics-explained/index.php/Glossary:Labour_force_survey_(LFS).

2.5 Confidentiality and Access to Confidential Data

Confidentiality of statistical data is a principle that is of immense strategic importance to official statistics. Only when it is completely certain that respondents' information will be treated with complete confidentiality and that it will not be used for any purpose other than for statistical purposes can truth-based answers and statements in statistical interviews be expected. Any suspicion of other uses—for tax or other administrative purposes, not to mention police or judicial investigations—undermines the delicate relationship of trust between citizens, businesses and the (public) statistical institution.

In this sense, the judgment of the German Constitutional Court on the planned census of 1983 (openJur 1983) is to be interpreted. The court emphasised the right to 'informational self-determination' and prohibited a return of statistically collected data to the administration for the purpose of correcting errors in the registration records. Since then, statistics have been regarded as a closed-off area into which administrative data can enter, but not return back.

A development in the opposite direction has increased massively since the 1990s. For research, especially on social science issues, the aggregated results, as provided in tables and databases of official statistics, are important. However, the actual wealth of the micro-data is insufficiently explored. It was therefore necessary to open up ways and means of access to individual data for researchers, without endangering confidentiality. Through a combination of legal requirements with new mathematical methods, as well as through special secluded premises, such access was realised (under restrictive conditions, after all).

Among other things, the intensive discussion of this topic has led to closer cooperation between statisticians and researchers, as well as the establishment of new working groups and committees.[42]

2.6 Modernisation

A widely supported starting point concerning the strategic orientation for official statistics is:

> Our output has traditionally been determined by the demands of our respective governments and other customers. The process is one of reasoning back from the output desired to survey design because often few or no pre-existing data were available. This paradigm has shaped the way official statistics are designed and produced. … In the future it will become increasingly unrealistic to expect meaningful statistics from this approach, even when results are collected and transmitted electronically. (Vale 2017)

Since the end of the 1990s, a re-engineering of the business model has been ongoing, according to which the single statistical production lines are bundled and

---

[42] Such as the German Data Forum RatSWD (https://www.ratswd.de/en/ratswd/development).

integrated, common technical tools are developed and terminology is standardised, thus minimising redundancies, inefficiencies and sources of incoherence. Information is generated by (re-)using available data as far as possible, aiming at minimising the response burden and costly surveys. In terms of the above-mentioned learning process, this means that the current 'development loop' is driven by changes on the production side, which will lead to substantial improvements.

Nevertheless, one must take the implications for the other actors of the learning cycle into account. For example, it was not difficult in the past and with the traditional business model to organise functioning user-producer dialogues, since participants of these dialogues shared the interest and knowledge of same subject matter. Agricultural statistics was discussed between the specialists for agricultural policies and the technicians in a specialised branch of the statistical office; the same applied for labour market, population, health statistics and so forth; a balanced agreement sufficient for static and narrow user needs. As long as statistical offices did not have to cope with substantial resource scarcities (and rapidly changing user needs), it was therefore not necessary to establish an overall programme planning, to decide on priorities, etc. The programme was just the sum of a great number of partial solutions in each separated area; both users and producers were generally satisfied—users with their tailor-made products and producers with their control of the entire production process. This inefficient 'spaghetti-bowl-business-model' of the past is replaced the new 'industrialised-process-model': multiple-source inputs, standardised production and multiple purposes output. The new business model of production cannot be 'administrated' in a traditional manner. It needs to be 'managed', including the development of planning tools, a catalogue of products/services, marketing and cost accounting, which means not less than a complete overhaul of the traditional culture in official statistics.

The national and international modernisation programmes are meanwhile characterised by great convergence. From a European point of view, the 'Vision 2020' of the European Statistical System (Eurostat 2018a) as well as the cooperation in the United Nations Economic Commission for Europe (UNECE) (2018) deserves special mention.

At this point, therefore, a summary of the essential components should suffice.[43] Work in this area includes:

- Production and IT: development and introduction of a common business architecture as reference framework for the production processes, improving the conditions for the sharing of IT services and infrastructure and the exchange of (micro)data
- Data sources and data collection: development of methodologies for mixed-mode and multisource collection, concepts for risk management in using new tools and sources
- Communication: development of a strategy for dissemination and communication, operational and innovative communication tools
- Standards and metadata: development of a metadata glossary, standards for linked open data/metadata

---

[43]For further insight see https://www.unece.org/stats/mos/meth.html.

- Human resources and organisational frameworks: creating positive conditions and capacities for the change management needed, training and learning, performance management, building competencies, introducing cost accounting for products and modernisation projects.

In fact, compared to the world of official statistics some 25 years ago, everything has fundamentally changed. Instead of a highly fragmented, de-centrally driven production, a centrally managed manufacturing, based on modular components, has entered. After a first phase, where registers and administrative sources have been established and used for statistical purposes, large and expensive surveys (such as the censuses) have been replaced more and more by mixed procedures, saving the costs and the burden on respondents. The only prominent case, for which a change went into the opposite direction (replacing an administrative data-based statistic to a survey-based statistic) is Intrastat, the statistics on the trade in goods between countries of the EU.

Nonetheless, this statistical modernisation, like any form of industrialisation, has its price. The tailored shoe and suit from the tailor fit better than the clothing from the factory. However, the key question is what you can afford and what is 'Fit for Purpose' under these conditions. The answer to this question is clear: it is only with the considerable efforts and efficiency improvements of recent decades that it is possible for official statistics to meet the political and social demands of today.

2.7 Conclusion: Official Statistics—Modern, Efficient, High Quality

Before elaborating on the driving forces of science and of political conditions in the history of official statistics in the next chapter, the following can be summarised:

First, official statistics are the result of a scientific process. If this sounds too ambitious from a purely scientific position, it may at least be pointed to scientific methods as their basis. Official statistics can be seen as a sub-category of 'scientific data and information' that helps to better understand how societies function and develop.

Secondly, it is equally important that the categories and variables used in official statistics reflect and represent societal conventions (Desrosières 2010, p. 126) and: '*To surpass the great divide between knowledge and politics means to take the tools of knowledge seriously politically*' (Desrosières 2010, p 127). Official statistics are similar in this respect to the legal system; legislative rules are initially defined as a convention before they are subsequently executed by the administration (Supiot 2015; Radermacher and Bischoff 2018 forthcoming).

Third, official statistics is a factory whose task is to guarantee the regular production of standardised information and related services. This 'industrial' production has its strengths as well as its limitations. For example, large-scale surveys or even censuses can be pursued and regularly repeated under these conditions. On the other hand, new developments and innovations are to carefully assess whether they can finally enter routine operations and meet high-quality standards; budgetary restrictions require the setting of priorities in the statistics programme, etc. As in companies in industry, the application of quality management in all its facets (such as EFQM) is now commonplace in official statistics. The biennial quality conferences in European statistics[44] are part of this and give a broad overview.

Fourth, as a consequence, the statistical information produced by official statistics are first and foremost 'products'. Similar to all other products, they serve a purpose and their quality should be fit for that purpose; a simple conclusion with far-reaching implications. Mainly, it means that quality cannot be seen as something absolute or purely scientific, like under laboratory conditions (Hoof et al. 2014), rather than depending on many reality factors: *'Quality by Design'* (Juran 1992); *"What, then, should be taken as priority: data utility, or rather, data quality? Such dilemmas have no preconceived wrong or correct answers. It is best to balance both sides of a conflict in a sensible way."* (Piotrowska-Piątek and Sobieraj 2016, p. 20)

Fourth, official statistics provide a public infrastructure (a public good). It is therefore part of the state administration, which also creates special framework conditions for working conditions. As a public infrastructure, official statistics are basically taxfinanced; their products and services are available free of charge to anyone; therefore, 'open data' is already taken for granted by official statistics. This has hitherto mainly applied to aggregated results of official statistics, while access to micro-data for legal reasons is subject to special regulations, just as the additional expense of providing micro-data to science requires separate funding. Here, too, solutions should be found which already take account of the current need for research data in the basic budgeting of official statistics.

Sixth, official statistics have much in common with the media, in particular, the public ones. The dissemination of information is the common mission and denominator. Close cooperation with journalists and an active role in social media are therefore just as essential for official statistics as keeping pace with the latest forms and methods of communication (Eurostat 2017a, c).

Finally, the way in which official statistics (i.e. statistical offices) are institutionalised and organised reflects the understanding of the role of the state as part of a political agenda. There is not one unique answer to the question *'Why Have Government Statistics? (And How to Cut their Cost)'* (Thomas 1984). It was a symbolic act that Margaret Thatcher first targeted the official statistics of the UK when implementing her neo-liberal program of shrinking the state (will say the public administration).[45] Similar to the public media, such as radio or TV broadcasters, it is necessary to regularly critically examine which part of the statistical information should be provided by public institutions and for which private providers (such as research institutes, opinion polls or universities) are better suited. As a rule, the answer to

---

[44]See the overview in http://www.q2018.pl/previous-q-conferences/; the first one was the quality conference 2001 in Stockholm, Sweden initiated by the LEG Quality [recommendation 14 (Lyberg et al. 2001)].

[45]see in particular "The legacy of Rayner" (Guy and Levitas 2005, p. 7).

this question has emerged from the evolution of a country's history in statistics and depends on whether the cooperation between public and private actors is effectively organised in practice. However, this also means that if conditions change (e.g., in the course of digitisation), adaptation and rationalisation of the different activities must be considered.

Since the beginning of the 1990, the environment around statistics has dramatically changed due to several factors, such as:

- Pressure on the public sector; major cuts in budgets and human resources
- Reduced willingness to respond to statistical surveys; response burden as a political target
- Exponentially growing importance of ICT and new data sources (e.g., administrative data, GIS)
- New political demands (e.g., environment, globalisation, migration) and crises (e.g., financial).

Official statistics has successfully met these challenges. With a radical adaptation of the business model, productivity was noticeably improved and new opportunities were created, even though budgetary conditions continued to deteriorate.

It will now be important to maintain this momentum and the change process. This will make it possible to make the transition to **Official Statistics 4.0** and to successfully master the associated challenges, above all from digitisation and globalisation.

References

Blanc, Michel, Walter Radermacher, and Thomas Körner. 2001. Quality and Users—Chapter of the Final Report of the LEG on Quality. In *Final Report of the LEG on Quality*, ed. Lars Lyberg, Mats Bergdahl, Michel Blanc, Max Booleman, Werner Grünewald, Marta Haworth, Lili Japec, Tim Jones, Thomas Körner, Håkan Lindén, Gunilla Lundholm, Margarida Madaleno, Walter Radermacher, Marina Signore, Ioannis Tzougas, and Richard van Brakel. Luxembourg: Eurostat.

Brandolini, Andrea. 2016. The Links Between Household Surveys and Macro Aggregate. In *DGINS 2016*, ed. Statistics Austria. Vienna.

CBS. 2016a. *Budget Cuts at Statistics Netherlands: State of Play*. Statistics Netherlands. Accessed January 29, 2018. https://www.cbs.nl/en-gb/about-us/organisation/operational-management/archief/budget-cuts-at-statistics-netherlands-state-of-play.

CBS. 2016b. *Statistics Netherlands Annual Report for 2015*. Hague: Statistics Netherlands.

Darabi, Anoush 2017. The UK's Next Census will be its Last—Here's Why. In *Apolitical*. London: Apolitical.

Davies. 2017. How Statistics Lost Their Power—and Why We Should Fear What Comes Next. *The Guardian*.

Deming, W.E. 2000. *Out of the Crisis*. Massachusetts Institute of Technology, Center for Advanced Engineering Study.

Deming Institute. 2018. *PDSA Cycle*. The Deming Institute, Accessed July 12, 2018. https://deming.org/explore/p-d-s-a.

Desrosières, Alain. 1998. *The Politics of Large Numbers—A History of Statistical Reasoning*. Cambridge, MA: Harvard University Press.

Desrosières, Alain. 2010. A Politics of Knowledge-tools—The Case of Statistics. In *Between Enlightenment and Disaster*, ed. Linda Sangolt. Brussels: P.I.E. Peter Lang.

Destatis. 2018. *Quality Assurance*. Destatis. Accessed January 29, 2018. https://www.destatis. de/EN/Methods/Quality/Quality.html.jsessionid=CFE078143E58C009209242ABF7DF5619. InternetLive2l.

ECB. 2018. *The ECB Statistics Quality Framework and Quality Assurance Procedures*. ECB. Accessed January 29, 2018. https://www.ecb.europa.eu/stats/ecb_statistics/governance_and_quality_framework/html/ecb_statistics_quality_framework.en.html.

EFQM. 2013. *EFQM Excellence Model*. Brussels: EFQM.

ESGAB. 2018. *European Statistical Governance Advisory Board*. Eurostat, Accessed January 29, 2018. http://ec.europa.eu/eurostat/web/esgab/introduction.

European Commission. 2009. Communication from the Commission to the European Parliament and the Council on the Production Method of EU Statistics: A Vision for the Next Decade. In *COM(2009) 404*.

European Commission. 2010. *Report on Greek Government Deficit and Debt Statistics*. Brussels: European Commission.

European Commission. 2011. *Towards Robust Quality Management for European Statistics*, ed. Eurostat, 10. Brussels: European Commission.

European Union. 2011. European Statistical Programme 2013–2017. In *COM(2011) 928 final*, ed. the European Parliament and the Council. Brussels: European Commission.

European Union. 2013. *Regulation (EU) No 99/2013 of the European Parliament and of the Council of 15 January 2013 on the European Statistical Programme 2013–17*, ed. European Commission, 9 February 2013. Brussels: Official Journal of the European Union.

European Union. 2015. Regulation (EC) No 223/2009 of the European Parliament and of the Council of 11 March 2009 on European statistics and repealing Regulation (EC, Euratom) No 1101/2008 of the European Parliament and of the Council on the transmission of data subject to statistical confidentiality to the Statistical Office of the European Communities, Council Regulation (EC) No 322/97 on Community Statistics, and Council Decision 89/382/EEC, Euratom establishing a Committee on the Statistical Programmes of the European Communities. In *2009R0223—EN— 08.06.2015—001.001—1*, ed. European Union. Luxembourg: © European Union, https://eur-lex. europa.eu, 1998–2019.

Eurostat. 2009. *Communication from the Commission to the European Parliament and the Council on the Production Method of EU Statistics: A Vision for the Next Decade*, ed. European Commission. Brussels.

Eurostat. 2011. *European Statistics Code of Practice for the National and Community Statistical Authorities*—Adopted by the European Statistical System Committee 28th September 2011, ed. Eurostat. Luxembourg: Eurostat.

Eurostat. 2013a. *ESS Vision 2020—Building the Future of European Statistics*. Eurostat. http://ec. europa.eu/eurostat/web/ess/about-us/ess-vision-2020.

Eurostat. 2013b. *European System of Accounts ESA 2010*. Luxembourg: Publications Office of the European Union.

Eurostat. 2013c. *Regulation (EU) No 549/2013 of the European Parliament and of the Council of 21 May 2013 on the European system of national and regional accounts in the European Union*, ed. European Union. Brussels: Official Journal of the European Union.

Eurostat. 2015. Regulation (EC) No 223/2009 of the European Parliament and of the Council. In *2009R0223—EN—08.06.2015—001.001—1*. Luxembourg: Eurostat.

Eurostat. 2016a. *Report from the Commission to the European Parliament and the Council on the Implementation of Regulation (EU) No 691/2011 on European Environmental Economic Accounts*. Brussels: European Commission.

Eurostat. 2016b. *Sharing Statistical Production and Dissemination Services and Solutions in the European Statistical System*, ed. Dir B. Luxembourg: Eurostat.

Eurostat. 2017a. *Digicom—Users at the Forefront*. Eurostat. http://ec.europa.eu/eurostat/web/ess/ digicom.

Eurostat. 2017b. *Eurostat Catalogue of Statistical Products 2017*. Luxembourg: Eurostat.
Eurostat. 2017c. *PART 2—Communicating Through Indicators*. Luxembourg: Eurostat.
Eurostat. 2018a. *ESS Vision 2020*. Eurostat, Accessed January 30, 2018. http://ec.europa.eu/eurostat/web/ess/about-us/ess-vision-2020.
Eurostat. 2018b. *European Statistics Code of Practice—For the National Statistical Authorities and Eurostat*, 20. Luxembourg: Publications Office of the European Union.
Eurostat. 2018c. *Peer Reviews in the European Statistical System*. Eurostat. Accessed January 29, 2018. http://ec.europa.eu/eurostat/web/quality/peer-reviews.
Eurostat. 2018d. *Quality Assurance Framework of the European Statistical System*. Eurostat. http://ec.europa.eu/eurostat/documents/64157/4392716/ESS-QAF-V1-2final.pdf/bbf5970c-1adf-46c8-afc3-58ce177a0646.
Foucault, Michel. 1991. Governmentality. In *The Foucault Effect*, ed. Graham Burchell, Colin Gordon and Peter Miller. Chicago: Chicago University Press.
Guy, W., and R. Levitas. 2005. *Interpreting Official Statistics*. London: Taylor & Francis.
Hoof, F., E.M. Jung, and U. Salaschek. 2014. *Jenseits des Labors: Transformationen von Wissen zwischen Entstehungs- und Anwendungskontext* (transcript Verlag).
IFAC. 2014. *International Framework: Good Governance in the Public Sector—Executive Summary*. London: The International Federation of Accountants (IFAC) and the Chartered Institute of Public Finance and Accountancy (CIPFA).
IMF. 2016. *Statement by the IMF Executive Board on Argentina*. Washington DC: IMF.
IMF. 2017. *IMF Standards for Data Dissemination*. IMF. Accessed January 29, 2018. https://www.imf.org/en/About/Factsheets/Sheets/2016/07/27/15/45/Standards-for-Data-Dissemination.
IMF. 2018. *Standards & Codes*. IMF. Accessed January 29, 2018. http://www.imf.org/external/standards/index.htm.
INE. 2018. *Quality in Statistics*. Statistics Portugal. Accessed January 30, 2018. https://www.ine.pt/xportal/xmain?xpid=INE&xpgid=ine_qualidade.
ISI. 2010. *ISI Declaration on Professional Ethics*. The Hague The Netherlands: International Statistical Institute.
ISI. 2018. *ISI Declaration on Professional Ethics—Adopted in August 1985—Background Note*. ISI. Accessed January 29, 2018. https://www.isi-web.org/index.php/news-from-isi/151-ethics1985.
ISTAT. 2018. *Quality Commitment*. ISTAT. Accessed January 30, 2018. https://www.istat.it/en/about-istat/quality.
Klein, Julie Thompson. 2004. Interdisciplinarity and Complexity: An Evolving Relationship. *E:CO*, Special Double Issue Vol. 6: 2–10.
Juran, J.M. 1992. *Juran on Quality by Design: The New Steps for Planning Quality Into Goods and Services*. Free Press.
Kyi, Gregor, Bettina Knauth, and Walter Radermacher. 2012. A Census is a Census is a Census? In *UNECE-Eurostat Expert Group Meeting on Censuses Using Registers*. Geneva: UNECE Conference of European Statisticians.
Lyberg, Lars, Mats Bergdahl, Michel Blanc, Max Booleman, Werner Grünewald, Marta Haworth, Lili Japec, Tim Jones, Thomas Körner, Håkan Lindén, Gunilla Lundholm, Margarida Madaleno, Walter Radermacher, Marina Signore, Ioannis Tzougas, and Richard van Brakel. 2001. *Summary Report from the Leadership Group (LEG) on Quality*. Luxembourg: Eurostat.
OECD. 2015. *Recommendation of the OECD Council on Good Statistical Practice*, ed. Statistics. Paris: OECD.
openJur. 1983. *BVerfG · Urteil vom 15. Dezember 1983 · Az. 1 BvR 209/83, 1 BvR 484/83, 1 BvR 420/83, 1 BvR 362/83, 1 BvR 269/83, 1 BvR 440/83 (Volkszählungsurteil)*. Accessed February 26, 2018. https://openjur.de/u/268440.html.
Piotrowska-Piątek, Agnieszka, and Małgorzata Sobieraj. 2016. On the Way to the Statistics' Quality. Quality as a Challenge of Public Statistics in Poland. *World Scientific News* 48: 17–23.
Porter, M.E. 1995. *Trust in Numbers: The Pursuit of Objectivity in Science and Public Life*. Princeton, N.J., Chichester: Princeton University Press.

Radermacher, Walter. 1992. Methoden und Möglichkeiten der Qualitätsbeurteilung von statistischen Informationen aus der Fernerkundung/Methods and Possibilities of Assessing the Quality of Statistical Data of Remote Sensing. *Jahrbücher für Nationalökonomie und Statistik*, 169–79.

Radermacher, W. 1999. Indicators, Green Accounting and Environment Statistics: Information Requirements for Sustainable Development. *International Statistical Review: A Journal of the International Statistical Institute and Its Associations* 67: 339–354.

Radermacher, Walter. 2008. Data Quality in Multiple Source Mixed Mode Designs. In *Q2008 European Conference on Quality in Official Statistics*, Rome.

Radermacher, Walter. 2011. ESS Vision and Ways for Cooperation. In *Workshop on Strategic Developments in Business Architecture in Statistics*, ed. UNECE. Geneva: UNECE.

Radermacher, Walter J. 2012. Zahlen zählen—Gedanken zur Zukunft der amtlichen Statistik in Europa. *AStA Wirtsch Sozialstat Arch* 2012: 285–298.

Radermacher, Walter. 2014. The European Statistics Code of Practice as a Pillar to Strengthen Public Trust and Enhance Quality in Official Statistics. *Journal of the Statistical and Social Inquiry Society of Ireland* 43: 27–33.

Radermacher, Walter and Pierre Bischoff. 2018 forthcoming. Article 338. In *Treaty on the Functioning of the European Union—A Commentary*, ed. H.-J. Blanke, and S. Mangiameli. Springer.

Radermacher, Walter, Joachim Weisbrod, and Dominik Asef. 2004. Bedarf, Qualität, Belastung Optimierung als Interessenausgleich. In *WiSta Wirtschaft und Statistik*.

Saetnan, Ann Rudinow, Heidi Mork Lomell, and Svein Hammer. 2011. *The Mutual Construction of Statistics and Society*. New York, NY: Routledge.

Stamhuis, Ida H. 2008. Statistical Thought and Practice. A Unique Approach in the history and development of sciences? In *The Statistical Mind in Modern Society. The Netherlands 1850–1940*, ed. I.H. Stamhuis, P.M.M. Klep, and J.G.S.J. van Maarseveen. Amsterdam: Aksant.

Supiot, Alain. 2015. *La Gouvernance par les nombres*. Nantes: Librairie Arthème Fayard.

Thomas, Ray. 1984. Why Have Government Statistics? (And How to Cut their Cost). *Journal of Public Policy* 4: 85–102.

UNECE. 2007. *Register-Based Statistics in the Nordic Countries—Review of Best Practices with Focus on Population and Social Statistics*. Geneva: United Nations Publication.

UNECE. 2013. *Generic Statistical Business Process Model (GSBPM)*. http://www1.unece.org/stat/platform/display/GSBPM/Generic+Statistical+Business+Process+Model.

UNECE. 2017. *The Generic Statistical Information Model (GSIM)*. UNECE. https://statswiki.unece.org/display/gsim/Generic+Statistical+Information+Model.

UNECE. 2018. *Modernization of Official Statistics*. UNECE. Accessed January 30, 2018. https://www.unece.org/stats/mos.html.

United Nations. 2014. *Fundamental Principles of Official Statistics*. New York.

UNSD. 2018. *Fundamental Principles of National Official Statistics*. UNSD. Accessed January 29, 2018. https://unstats.un.org/unsd/dnss/gp/fundprinciples.aspx.

Vale, Steven. 2017. *Strategic Vision of the HLG-MOS. UNECE High-Level Group for the Modernisation of Official Statistics*. http://www1.unece.org/stat/platform/display/hlgbas/Strategic+vision+of+the+HLG-MOS#StrategicvisionoftheHLG-MOS-I.Introduction.

Chapter 3
Science and Society: A Reflexive Approach to Official Statistics

In this chapter, we open a large box with questions and reflections about the scientific background of official statistics. First, it will be about knowledge: how can we know that we know what we know (or do not know)? Then we will shed light on the social position, role and function of statistics (in the sense of science, information and institution). Some episodes from the history of official statistics are used for clarification. Finally, two concrete and current applications will conclude the chapter: Indicators and Sustainable Development. But first, a methodology is presented that provides a holistic roadmap through this extremely broad topic, the 'System of Profound Knowledge' by W. E. Deming.

3.1 Profound Knowledge—A System's Approach to Quality

"*'In God we trust. All others must use data' If there is a credo for statisticians, it is that*" (Walton 1986, p. 96). As with many other quotes about statistics, this one only tells a (very small) part of the approach for better management that W. E. Deming was teaching. Deming started his career as a statistician before becoming the founder of Total Quality Management (TQM). His views on statistical methods were twofold. First, he saw statistical methods as important tools "*to understand processes, to bring them under control, and then to improve them*". But he emphasised "*that to rely on statistical methods alone is not nearly enough*" (Walton 1986, p. 20). In his later book, *The New Economics* (Deming 2000), he developed his "*System of Profound Knowledge*" with four parts, all related to each other.[1] These four parts are system, variation, theory and psychology.

Appreciation for a system: understanding the overall processes involving suppliers, producers and customers (or recipients) of goods and services, understanding how interactions (i.e. feedback) between the elements of a system can result in internal restrictions that force

---

[1] See *Deming's System of Profound Knowledge* in https://blog.deming.org/2012/10/demings-system-of-profound-knowledge/.

© Springer Nature Switzerland AG 2020
W. J. Radermacher, *Official Statistics 4.0*,
https://doi.org/10.1007/978-3-030-31492-7_3

the system to behave as a single organism that automatically seeks a steady state. It is this steady state that determines the output of the system rather than the individual elements. Taking a systems approach results in management viewing the organisation in terms of many internal and external interrelated connections and interactions, as opposed to discrete and independent departments or processes governed by various chains of command. When all the connections and interactions are working together to accomplish a shared aim, a business can achieve tremendous results from improving the quality of its products and services, to raising the entire esprit de corps of a company.

Knowledge about variation: the range and causes of variation in quality and use of statistical sampling in measurements. In any business, there are always variations—between people, in output, in service and in product. The out of a system results from two types of variation: common cause and special cause variations. Common cause variations are the natural result of the system. In a stable system, common cause variation will be predictable within certain limits. Special cause variations represent a unique event that is outside the system; for example, a natural disaster. Distinguishing the difference between variations, as well as understanding its causes and predicting behaviour, is key to management's ability to properly remove problems or barriers in the system.

Theory of knowledge: the concepts explaining knowledge and the limits of what can be known. How do we know that what we think we know is really so? There is no true value of any characteristic, state or condition that is defined in terms of measurement or observation. The 'value' is in the context for a given operational definition. Understanding that a value must be interpreted via context leads us to question any data that does not provide the operational definition for how the data was created. And this leads to better understanding of whether or not the data is really useful, because without having the operational definition we are likely to draw incorrect conclusions from data.

Knowledge of psychology: concepts of human nature. An organisation has a duty to create a system where people can take pride in what they do. By doing so, the organisation is able to focus on continual customer-focused improvement over the long term. Deming's view is that employees are key to the long-term success of the organisation. They are not costs to be minimised; they are valuable partners in the continuing success of the organisation.

Deming's teaching on management was very often summarised in the form of principles that have been incorporated into the standard textbooks and TQM training courses. More practically, he condensed his philosophy in '14 Points for Management and Seven Deadly Diseases of Management', for example "*5. Management by use only of visible figures, with little or no consideration of figures that are unknown or unknowable*" (TheDemingInstitute 2018).

Interestingly, it turns out that Deming's intention was by no means the construction of a monstrous measurement machinery for controlling and accounting, rather the opposite: "*It is wrong to suppose that if you can't measure it, you can't manage it – a costly myth*" (Deming 2000, p. 35).

Dr. Deming did very much believe in the value of using data to help improve the management of the organization. But he also knew that it wasn't close to enough. There are many things that cannot be measured and still must be managed. And there are many things that cannot be measured and managers must still make decisions about. (Hunter 2015)

Neither did the other management guru Peter Drucker ever express "*If you can't measure it, you can't manage it*". Rather, "*Drucker's take on measurement was quite nuanced*", not measurement myopia (Zak 2013).

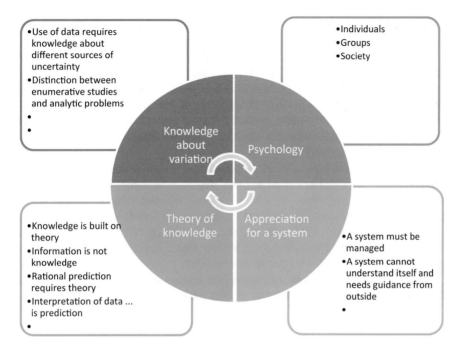

Fig. 3.1 Components of the system of profound knowledge. See the full composition of Deming's approach to Profound Knowledge here: Moen and Norman (2016: 50)

The approach used in this chapter follows Deming's concept. It is all about profound knowledge for official statistics. How the system of official statistics is structured and how it works have already been explained in the previous section. The point here is to understand the possibilities and limits of measurement and knowledge creation, and what Deming calls 'psychology' in our case, accordingly, the sociology of the interactions between statistics and society (Fig. 3.1).

Following the concepts of Russell Ackoff,[2] a systemic approach will be applied to firstly analyse official statistics in its individual parts and then to get an overview of the whole with its functions in a synthesis.

With such a broad and multidisciplinary approach, however, one runs the risk of completely losing oneself in the expanses and depths of other disciplines, such as the philosophy of knowledge or of sociology. One must therefore select, and priorities must be set. The selection criterion for the treatment of topics as well as the detail of their representation is their relevance to official statistics. 'Profound knowledge' does not mean universal knowledge. However, it is a walk on a fine line and the attempt to achieve the optimum between the goal of completeness and the accuracy of detail, even if these goals are in a certain conflict with each other.

---

[2] 'Systems thinking is the fusion of analysis and synthesis, depending on whether our objective is knowledge or understanding' (Ackoff 1994).

The second part of this chapter will focus on the most important epistemological questions behind official statistics. It is necessary to ask with what expectations one can and should approach statistics and the results produced by them. Are the reality and statistics that reflect reality one and the same? If there is a difference between reality and its (statistical) image, how does this image emerge, what can it provide and what can it not provide in terms of 'truth'? How should one deal in general with terminology such as truth, reality, correct, right and wrong? What is the appropriate manner of a statistician to communicate results without risking committing the fallacy of misplaced concreteness?[3]

The third will deal with scientific approaches that analyse the 'co-construction' phenomenon: *"Statistics are often seen as simple, straightforward, and objective descriptions of society. However, what we choose to count, what we choose not to count, who does the counting, and the categories and values we choose to apply when counting, matter"*.[4] This part will concretise the co-construction of statistics and society for selected historical episodes and present the current mainstream way of thinking as far as is relevant for official statistics.

The fourth and fifth parts will apply the lessons learned to a more general discussion of statistical indicators and metrics for Sustainable Development.

3.2 Epistemology—Theory of Knowledge

3.2.1 The Truth, Reality and Statistics

Every statistician (especially in official statistics) is confronted in the course of his or her professional life with the question of whether the presented statistical results are 'correct', whether they correspond to the 'truth' and reflect the 'reality'. Such questions express the natural need of users to be assured of the reliability and (more generally) the quality of the numbers produced.[5] Facts and figures are commonly regarded as the opposite of opinions. One even speaks of 'hard facts'. That is a good

---

[3]See definition in https://en.wikipedia.org/wiki/Reification_(fallacy)#Fallacy_of_misplaced_concreteness.

[4]See the introduction of (Saetnan et al. 2012) on the Routledge website https://www.routledge.com/The-Mutual-Construction-of-Statistics-and-Society/Saetnan-Lomell-Hammer/p/book/9780415873703.

[5]The great interest in this topic is, for example, clarified in the conference 'Truth in Numbers: The role of data in a world of fact, fiction and everything in between', 4 April 2018, Bern, Switzerland (http://www.paris21.org/news-center/events/conference-truth-numbers-role-data-world-fact-fiction-and-everything-between) or by Hans Rosling's book 'Factfulness' (Rosling et al. 2018) and corresponding reactions (e.g. Population Matters 2018).

thing, especially in times when this characteristic of the facts is under pressure and the truth is replaced by arbitrariness.[6]

It is, however, necessary to get to the bottom of the matter and clarify what it means to produce and use 'facts'. An unrefracting, badly prepared participation in this (public) debate, which is loaded with subjective value judgements and even emotions, can cause great damage to statistics. Trust could easily be lost because of misunderstandings and wrong perceptions or expectations. If one contextualises statistics with undefined terms and values, such as right and wrong, truth and false-hood, then there is a great danger that statistical illiteracy, mixed with a general scepticism against statistics (resulting from misunderstood or poorly communicated statistical information) and combined with a general mistrust against state institu-tions or actors, leads citizens to the erroneous conclusion that such facts provided by official statistics are just not harder than alternative low-quality figures or even their own opinion.[7]

Vague terminology is unsuitable for assessing the quality of statistics: to judge whether statistics are true and whether they correctly reflect reality would require an (absolutely true) reference measure of this reality, which, of course, does not exist (since this is exactly the role and function of the statistics under consideration).

There is a way out of this dilemma: 'Quality' has to be translated into concrete definitions and concepts that, on the one hand, build on scientific knowledge but on the other hand are also concrete, pragmatic and feasible, and their implementation can be reviewed. In principle, this is not difficult once one has accepted to take statistical information for what it is: a product. Using the definition of quality introduced in Chap. 2, **a statistic is of good quality if it has a good design, if it has been produced correctly according to the chosen design, statistical principles and standards, and if its results have reached the addressees with good and comprehensible communication**. If one of these conditions are not fulfilled, be it by inadequate design, be it by errors and mistakes in the production or be it by miscommunication, statistics have insufficient, bad quality.

In this way, inadequate, ambiguous and value-loaded contexts of 'truth', 'correct-ness', etc., are substituted by another conceptual frame: Total Quality Management. TQM provides an appropriate, concrete and feasible benchmark for the assessment of statistics. It includes all the ingredients that a methodological framework needs to produce and constantly improve quality. It addresses the producers of statistics as well as the stakeholders. Concrete criteria for quality are defined, and their implemen-tation is tested and certified. All this is transparent and comprehensible to external parties.[8]

---

[6]See, for example, Michiko Kakutani in *The Guardian*, 'The death of truth: how we gave up on facts and ended up with Trump' (https://www.theguardian.com/books/2018/jul/14/the-death-of-truth-how-we-gave-up-on-facts-and-ended-up-with-trump) (Kakutani 2018).

[7]William Davies has emphasised this major risk for the future of statistics (Davies 2017).

[8]In his elaboration on "*Science beyond truth and enlightenment?*" Ulrich Beck asks for a "*reflexive, complete scientization, which also extends scientific scepticism to the inherent foundations and consequences of science itself*" (Beck 1998, p. 155).

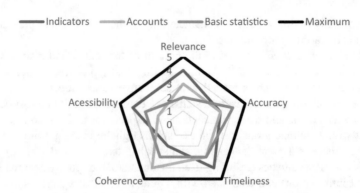

Fig. 3.2 Spider diagram for quality of statistical information. Adapted from Ravetz et al. (2018, p. 279)

Particularly, this approach to quality is specified in Europe by the ES Code of Practice presented in Chap. 2. The quality criteria of this Code of Practice (see Fig. 2.13) can be used to assess the quality of statistical information and to present the results of this assessment as a quality profile, for example, in form of a spider diagram, like in Fig. 3.2.

3.2.1.1   A Few Typical Examples that Could Illustrate This Question

First case: Are the results of the population census correct?
In a lawsuit, the Federal Constitutional Court of Germany deals with the complaint of some local authorities and municipalities, which cast doubt on the results of the 'Census 2011'. On the deadline of 9 May 2011, a nationwide population, building and housing census took place. This was for the first time a so-called register-based census. In contrast to the previous censuses, only about 10% of the population was surveyed. To reduce the number of required surveys, data already collected in registers has been used; these were supplemented by primary statistical surveys. Among other things, the census established the official population figures for all communities in Germany. In addition to a double counting test in municipalities with at least 10,000 inhabitants, the procedure provided for a random sample survey in order to correct inaccuracies in the register of migrants, as well as a so-called survey to clarify discrepancies in smaller municipalities. Among other things, the results led to the population of Berlin being determined to be around 180,000 lower than the population update, and the Hamburg population to be around 82,800 fewer. In particular, the applicants allege that the rules in the budget sample infringe certainty requirements. In addition, the statistical basis of the sampling procedure was not determined accurately enough. A sufficiently precise population investigation was not assured. The application of various methods depending on the size of the municipality was

incompatible with the requirements of inter-municipal and federal equal treatment, as it led to a disadvantage for larger communities (Bundesverfassungsgericht 2017).

This was not the first trial before the Federal Constitutional Court. Previously, in 1983, the census had been stopped by the court. At that time, it was essentially about the demarcation of the statistical sphere and the fundamental right to informational self-determination, which has laid one of the foundations for modern data protection (Bundesverfassungsgericht 1983).

The current lawsuit is well suited to tracking the questions of truth and reality in statistics. On the one hand, there is a lot of nostalgia, which transfigures the old-fashioned door-to-door census as an error-free and 'correct' methodology, because supposed counting is simple, robust and not burdened with a statistical-technical machinery. There is an anecdote from the earlier days, according to which a statistician should have responded to a judge's question as to how the correctness of the results was ensured: 'Your Honour, we have re-counted the results'. The judge was convinced and satisfied with the argument. Even if this story did not happen exactly as reported here, it could nevertheless identify the widespread, almost naive idea that one could simply count the population of a whole country as a herd of sheep. Since the days of Emperor Augustus,[9] however, a traditional census has been a mammoth undertaking, solving a variety of technical and methodological problems, using estimation techniques and making inaccuracies and mistakes unavoidable. Even if it does not appear so to non-statisticians, the questions of how accurately and undistorted a statistic reflects the reality are essentially the same, regardless of whether a traditional census is used or a multiple source approach, which is based on other existing information and sampling techniques, thus being able to reduce costs in this way. What remains are subjective question marks, impressions and reservations, which seem to be more serious regarding modern statistical methods (Bubrowski 2017; Davies 2017). As the German court case and the comments in the public show, it is essential that such impressions and questions are taken seriously and that they are already eliminated, at least prevented during the preparatory work (as far as this is possible, of course).[10]

The **lessons learned** are not related to the production of these statistics, but especially to the design and also to communication. When designing methodically and technically demanding statistics with far-reaching political consequences, as in this case of the census, it is crucial to leave absolutely no doubt that the design (of course, in the context of the exogenously given conditions) meets the current scientific standard. Transparency (explicit and detectable) sound scientific advice and active public relations work are crucial in a modern statistical process in order to familiarise a critical public with the procedures and results.

---

[9]*In those days Caesar Augustus issued a decree that a census should be taken of the entire Roman world.[2] (This was the first census that took place while Quirinius was governor of Syria.) And everyone went to their own town to register* (Luke 2 NIV, 2).

[10]The Federal Constitutional Court announced its judgment in this case on 19 September 2018 (https://www.bundesverfassungsgericht.de/SharedDocs/Pressemitteilungen/EN/2018/bvg18-074.html).

In any case, when there is criticism of supposedly incorrect results, we should at first focus on the concrete and answerable questions about the adequacy of the design of statistics and its error-free implementation.[11]

Second case: Have inequalities in our societies worsened?
Ever since Thomas Piketty presented his analysis of long-term developments in the distribution of income and wealth (Piketty 2014), concern for and interest in this subject have grown.

> The statistical coverage of the material aspects of inequalities is linked to the distributions of income, consumption and wealth. There are however some gaps in the official statistics in this respect. The first one concerns the joint distribution of income, consumption and wealth, at the individual level. Better use of existing instruments and progress towards multisource environment, linked to new matching techniques, could improve the situation. The second main gap is with the conceptual and data discrepancies between micro and macro data and frameworks. An approach based on the identification of similarities, integration and quality assessment is proposed so as to bridge between the two sources, without denying their value in their own sphere. (Gueye 2016)

The design of politics is difficult, if not impossible, as long as the statistical bases allow very different conclusions (Hufe et al. 2018). In this respect, it is understandable that the expectation, indeed the demand, on the part of the users in the media and politics is that such gaps are closed and contradictions in the statistics are eliminated.

What, however, is the 'true' distribution of income, consumption and wealth (ICW) right now? This issue was the subject of the Conference of the Directors General of the National Statistical Institutes (DGINS) in 2016.[12] European (Brandolini 2016) and international (Seneviratne 2016) expertise was presented and discussed, and a final memorandum decided (DGINS 2016).

We can **learn two lessons** from this characteristic case: The first is that the outcome of statistics is essentially determined by their design and their methodological input factors, such as data sources and data collection methods. It also plays a big role in whether a statistic is designed for the undistorted and error-free collection of data for a suitable subarea (a so-called 'micro' approach) or for the most consistent and complete mapping of the entire area, tolerating inaccuracies in detail (a so-called 'macro' approach). The second lesson is that users of statistics have very little understanding of such sophisticated argumentation lines from statistics. Especially in times of evidence-based decision-making, users are under considerable pressure to justify their conclusions with so-called hard facts. This results in the logic of the list of measures in the Vienna Memorandum and in the first place: '*to develop a harmonised ICW statistical framework. Using a multisource approach, a conceptual framework of standards and methods for European ICW statistics will be developed, also improving micro data coverage and micro – macro links for ICW statistics*' (DGINS 2016, p. 2). Comparable situations and discussions exist in many areas of

---

[11]See in particular the considerations and guidelines concerning the Zensus 2011 court case (Bundesverfassungsgericht 2018).

[12]http://www.statistik.at/dgins2016/.

official statistics, especially where interests of different social groups or decisions with financial implications are involved, e.g. in the labour market statistics (Eurostat 2017d; Clouet 2015).

But the question for further epistemological thoughts below is: How far will methodical harmonisation and standardisation ultimately succeed? Is there such a thing as a 'right' income distribution statistic or, just to mention a similar subject, 'correct' unemployment figures? Can statistical numbers be 'true' in the sense of being the opposite of 'false'?

Third case: Are Greek statistics on GDP and public finances correct?
Greek public deficit figures for 2009 (and all preceding years) were problematic, with the forecasts of the deficit to GDP ratio (which were prepared by the Ministry of Finance in 2009) having to be increased from an initial 3.7% to a final 12.5%. In April 2010, the first statistical estimate of the actual outcome increased the ratio to 13.6% but still Eurostat had concerns with the methodology used and published the statistics with reservations. Ultimately, and following a rigorous examination, a revised estimate of 15.4% was submitted to and published by Eurostat in November 2010.

In 2009, it was already the second time that Greece caused an earthquake in European statistics. Already in 2005 (one election of the Hellenic Parliament and a change of government earlier), misreporting and subsequent major/implausible revisions of macro-economic indicators had caused a crisis, which asked for a profound revision of the making of official statistics in Europe.[13] Unfortunately, all these new safeguards of quality did not prevent the misreporting from continuing until 2009, when (again after an election) the (next) Greek government revised the previously notified figures. What happened then[14] is summarised in the corresponding report of the European Commission of 8 January 2010 (European Commission 2010).

What is of interest here is the response from different parts of the public and the participating and interested actors. The international feedback on the revision of the numbers and the institutional changes was very positive; capital markets calmed down and international confidence in the credibility of Greek statistics could be slowly rebuilt. In this process of recovery of trust, it was crucial not only to inform users about the revised results, but also to give them a realistic understanding of the remaining uncertainty margins. Not surprisingly, the question that was asked regularly was when we could finally expect 'correct' results. To answer this professionally, it was pointed out that the benchmark for the quality of Greek numbers should be the average of the results of the other EU member states, which are also subject to minor inaccuracies within a tolerance interval (European Commission 2018).

The reactions within Greece were and still are in stark contrast. The statistics computed according to international standards are held liable for the fiscal programmes

---

[13]The famous Goodhart's law (https://en.wikipedia.org/wiki/Goodhart%27s_law) was proven in a dramatic reality check in Greece, after European macro-financial indicators were directly related to the Euro currency, the Stability and Growth Pact and even the Treaties themselves.

[14]See also '*A Greek Tragedy: Hubris, Ate, and Nemesis*' (Coyle 2015: 77).

and austerity obligations since 2010. Paradoxically, 'falsification'[15] is alleged where the statistics administration acted for the first time neutrally and impartially. As a dramatic consequence of this intra-Greek conspiracy allegation, the responsible professional and administrator for these statistics at the top of ELSTAT, the statistical agency of Greece, Andreas Georgiou (as well as two other ELSTAT managers), was confronted with trials and sentenced. That this led to an overwhelming protest of all international statistical organisations[16] and national statistical societies has, so far, changed nothing.

Nothing can better illustrate where the exaggerated expectations and pitfalls are, when statistics and truth are mixed in a confused manner, like this dreadful case. Also, this case shows how much confidence in state authority or the absence of it leads one to trust official statistics to perform their task neutrally, impartially and purely technically scientifically (or not). If the administration is largely politicised, a citizen simply cannot imagine that unpolitical official statistics could possibly exist.

The more that value-loaded and normative terms such as 'truth' appear in a debate, the more statisticians should be mindful of following the rules of Deming's profound knowledge and communicating on their basis.

The **lessons learned** from this third case are—contrary to the two other ones—oriented towards the production process. In other words, the design of the statistics in question, namely the calculation of public sector debt/deficit and National Accounts, is not up for disposition at national level. One has to acknowledge that official statistics in the EU is a rules-based system that ensures comparability and consistency in the application of statistical methodology throughout all EU Member States. There is no such thing as scientific 'freedom of choice' to apply whatever kind of methodology in the core of European statistics, which is based on standard methods and classifications that are manifested in European legislation, as adopted by European Council and European Parliament. All Member States (and consequently all public statistical institutes) have to stick to these rules (that Member States as the main legislators in the EU have decided themselves). In this governance framework, Eurostat is the final statistical authority and guardian of the Treaties, thus ensuring that the rule of law is applied equally throughout the whole EU.

However, if the design is already regulated by default, it is all about producing statistics exactly to that standard. However, this has not happened in Greece, at least not until the year 2010. If one talks about false or fake Greek statistics, then what is meant is negligent errors from the statutory rules, weak statistical systems or even deliberate misreporting:

> Two different but in some instances linked sets of problems: problems related to statistical weaknesses and problems related to failures of the relevant Greek institutions in a broad sense. The first set of problems concerns methodological weaknesses and unsatisfactory

---

[15]Significantly, this allegation was made by a so-called Truth Committee (TruthCommittee 2015: 18).

[16]See for example the letters of AMStat (http://www.amstat.org/asa/files/pdfs/POL-20170901GeorgiouSeptember.pdf), FENStatS (http://fenstats.eu/data/news/Letter_FENStatS.pdf) or ISI (https://isi-web.org/index.php/activities/professional-ethics/isi-statements-letters).

technical procedures in the Greek statistical institute (NSSG) and in the several other services that provide data and information to the NSSG, in particular the General Accounting Office (GAO) and the Ministry of Finance (MOF). The second set of problems results from inappropriate governance, with poor cooperation and lack of clear responsibilities between several Greek institutions and services responsible for the EDP notifications, diffuse personal responsibilities, ambiguous empowerment of officials, absence of written instruction and documentation, which leave the quality of fiscal statistics subject to political pressures and electoral cycle. (European Commission 2010, p. 4)

Again, it becomes clear how tight the interlocking of official statistics and political action is. It is therefore all the more important to ensure with sound governance the independence and strength of the statistical institutions.

Official statistics must be policy-relevant but must not be politically driven.

3.2.1.2 Realism and Relativism

What is it that characterises the relation of statistics to reality and truth? The answer to this question leads to fundamental disputes between positions and schools of thought in the philosophical and sociological sciences. Can our knowledge and understanding of reality be considered objective or does it depend on the construction of the models we need in order to form a picture of natural or social phenomena? To what extent does science as a whole carry us with its findings and where are the limits? (Latour 1987; Benessia et al. 2016)

> The issues around knowledge – what we can know about the world, how we know it, what the status of our experiences is – have been central to philosophical reflection for ages. Answers to these questions, admittedly oversimplified here, have traditionally taken one of two forms. On the one hand there is the belief that the world can be made rationally transparent, that with enough hard work knowledge about the world can be made objective. … On the other hand, there is the belief that knowledge is only possible from a personal or cultural-specific perspective, and that it can therefore never be objective or universal. (Cilliers 2000, p. 8)

Barbieri uses a reference to the European statistical law as introduction and background to relate this question to official statistics:

> The European statistical law … devoted to the subject matter and scope of the Regulation itself, and to the relevant definitions and principles. Among the latter … one finds reliability, 'meaning that statistics must measure as faithfully, accurately and consistently as possible the reality that they are designed to represent and implying that scientific criteria are used for the selection of sources, methods and procedures.' … the rationale for this choice of the European legislator is clear enough: try and avoid putting statistics in the same class as the many 'narratives' competing for the attention of the public. (Barbieri 2018, forthcoming)

However, the ambiguity that eludes the legal basis for a precise definition of reality and the relation of statistics to it is common practice.

> This 'reality' is understood to be self-evident: statistics must 'reflect reality' or 'approximate reality as closely as possible'. But these two expressions are not synonymous. The very notion of 'reflection' implies an intrinsic difference between an object and its statistics. In contrast, the concept of 'approximation' reduces the issue to the problem of 'bias' or

'measurement error.' Thus even these two common expressions generally used without regard to consequences, tell us something important: a critical re-examination of this notion of 'reality' is for statisticians an efficient way to reconsider the deepest-rooted but also the most implicit aspects of their daily work.... (Desrosières 2001, p. 339)

Desrosières distinguishes two basic attitudes to reality[17]:

- 'Realism': *"The object to be measured is just as real as a physical object, such as the height of a mountain. The vocabulary used is that of reliability: accuracy, precision, bias, measurement error, ... this terminology and methodology was developed by the eighteenth century astronomers and mathematicians, ... The core assumption is the existence of a reality that may be invisible but is permanent – even if measurement varies over time. Above all, this reality is independent of the observation apparatus"* (Desrosières 2001, p. 341).

The difficulty with realism, however, is that *"'ultimate reality' is never accessible directly but only through different perception systems ... The realisms come together in a single test – that of the consistency between the various perceptions"* (Desrosières 2001, p. 349).

An entirely different approach is characterised in its concern to reconstruct the chain of coding and measurement conventions, thus effectively challenging the reality of the objects.

- 'Relativism': The explicit admission that the definition and coding of the measured variables are 'constructed', conventional, and arrived at through negotiation.

Although Desrosières' application example of business statistics has some peculiarities, his observation can be generalised. For example, Deborah Lupton comes to a closely related although more differentiated classification (Fig. 3.3).

With this juxtaposition of realism and relativism, a dilemma becomes clear that statisticians have to deal with:

Realism and relativism represent two polarised perspectives on a continuum between objective reality at one end and multiple realities on the other. Both positions are problematic for qualitative research. Adopting a realist position ignores the way the researcher constructs interpretations of the findings and assumes that what is reported is a true and faithful interpretation of a knowable and independent reality. Relativism leads to the conclusion that nothing can ever be known for definite, that there are multiple realities, none having precedence over the other in terms of claims to represent the truth about social phenomena. (Andrews 2012)

Of course, it is not the case that this dilemma is of great importance in the daily work of official statistics. Once the design and the measurement regime are decided for one statistic, i.e. when it is known which nomenclature is used, which population is included in the survey, how the sample is drawn, etc., then a technical–methodological orientation is in principle sufficient for the conduct of a high-quality production process. Taking Desrosières' example from business statistics, it becomes evident,

---

[17]In the attitude of 'realism', different ways of plausibility checks (i.e. ways of verifying and articulating the substance of that reality and its independence from observation) exist in different statistical communities, such as survey statisticians or accountants (Desrosières 2001).

Epistemological position	Key questions
Naïve realism: Reality is an objective phenomenon that exists and can be measured independently of social and cultural processes. Perceptions of reality may be distorted or biased through social and cultural frameworks of interpretation	What realities exist? How should one measure and manage them? How should information about realities be effectively communicated to the public? How to reduce 'bias' in the responses? How do people respond to questionnaires? What worldviews shape their responses?
Critical realism: Reality is an objective phenomenon, the measurement of which is inevitably mediated through social and cultural processes and can never be known in isolation from these processes	What is the relationship of reality and the measurement of reality to the structures and processes of 'late modernity'[84]
Relativism: Nothing is a reality in itself – what we understand to be a 'reality' is the product of historically, socially and culturally contingent 'ways of seeing'	How do the discourses and practices around reality operate in the construction of subjectivity, embodiment and social relations? How does reality operate as part of governmental strategies and rationalities?[85]

Fig. 3.3 Epistemological approaches in social sciences. Adapted from Lupton (2013, pp. 49–50)

however, that this technical orientation and realism in statistical production is shaped by the diverse cultures that have emerged in the different domains of statistics due to the methodological conditions and the interaction with scientists and stakeholders in each area. So how data collected is checked for plausibility, how the production process is controlled, how errors are discovered and eliminated and what is understood as good quality (i.e. the quality profile) is very specific to the individual statistical domains. In summary, it is noteworthy that evidently different forms and expressions of such 'realism' exist side by side. While the 'insiders' of a statistical domain (whether producers or users) focus on the technical–methodological issues and pay little attention to epistemological issues, their influence and importance become clear as soon as one looks at different areas from the outside and compares area-specific professional routines and cultures.[18]

Whenever new information needs to be poured into new statistical form or when the design of existing statistics is changed, when statisticians are faced with "*situations marked by controversy, crisis, innovation, and changes in economic, social and administrative contexts*," (Desrosières 2001, p. 349), decisions must be taken that ultimately require awareness and profound knowledge of the epistemological issues mentioned. In the design process (as well as in communication), it is part of the statistician's professionalism to be aware of the limitations of measurability, to reflect on the impact of statistics on society and to develop a basic understanding of complexity (and the role of statistics).

---

[18] '*More than one solution is possible because more than one measurement regime is possible, and this means that there is a range of potentially valid measures*' (Porter 1995, p. 33).

Therefore, it is imperative for a student or a researcher of science to differentiate between the computational tool and what it computes, to distinguish the map from the territory it represents. 'The map is not the territory', remarked Alfred Korzybski. There are multitudes of maps that we use to 'represent' the reality out there. They differ both in form and substance. The scientist in this sense resembles a cartographer. Only a cartographer knows how hard it is to represent a map of the earth on a sheet of paper. Every step towards perfecting the map involves a sacrifice – adding some feature to the map that does not have any intuitive or direct correspondence with the territory or ignoring many complexities of the territory. (Wuppuluri and Doria 2018, p. vii)

In the following, a middle course between realism and relativism is chosen (i.e. critical realism), on the one hand recognising a reality that exists independently of our perception, on the other emphasising that direct access to this reality is not possible, but requires methods of quantification, which inevitably contain simplifications, decisions and conventions.

Proper terminology
The decision for such a middle course should also be recognisable by dealing with terminology consciously and wisely.

There are erroneous (wrong) statistics that show evident blunders. Such 'errors' are primarily production errors in which something happened in the process that should not have happened. Of course, such manufacturing deficiencies must be distinguished from the fact that statistical results are subject to uncertainties that lie in the nature of the survey design and are planned a priori (random sampling error, etc.); these are therefore characteristics of the quality profile.

Likewise, there may be design or communication deficiencies that, however, make the use of the term 'error' more difficult: what about sample design that does not meet the scientific requirements of the state of art? Or is it a 'mistake' when we talk about 'foreign trade' when international trade in goods is meant? But what do you call a statistic that is ideally free from all these shortcomings? Is this statistic 'correct' or even 'true'? Although, as explained in the beginning of this section, the pressure of expectation is high that such attributes are used, we should beware of it and instead communicate things as they are.[19]

The term 'information' is used deliberately when it comes to statistics outputs.[20] Without elaborating on the meaning and definition of information at this point, reference is made to the introduction by Küppers:

Information is based upon symbols and sequences of symbols. ... we can distinguish three dimensions of information. The **syntactic** dimension denotes the way in which the symbols are arranged, as well as their relationship one to another. The **semantic** dimension includes the relationship among symbols and what they mean. Finally, the **pragmatic** dimension includes the relationship between the symbols, what they mean, and the effect that they engender with the recipient. (Küppers 2018, p. 68)

---

[19] *'For such a model there is no need to ask the question 'Is the model true?' If 'truth' is to be the 'whole truth' the answer must be 'No'. The only question of interest is 'Is the model illuminating and useful?'* (Box 1976, p. 792).

[20] Whereas the term 'data' is logically placed on the input side of the statistical process; unfortunately, 'data' is nowadays often used as a buzzword for the entire area of data, information and knowledge.

Statistical information is a product that is designed, produced and 'sold'. Such products are not wrong or right. In the best case, they meet pre-established and openly communicated minimum standards, such as methodological or ethical standards (a guarantee for users), and they must be measured against international best practice (also openly visible). In addition, the crucial question is whether the product portfolio as a whole provides an adequate (caution: not 'right'!) answer to the question of social progress, Sustainable Development and so on. "*Adequate measurement, clearly, means disciplining people as well as standardizing instruments and process*" (Porter 1995, p. 28).

3.2.2 Measurability, Models, Learning

The previous section basically explained the relationship between statistics and what is called 'reality'. The point now is to explain this process, in which reality is first depicted in an act of abstraction in a theoretical or qualitative model (where it is not important at this point whether this happens in language or mathematical formulas), then is translated into an empirical method of quantification. At this point, the concept of the 'model' as epistemic tool is introduced, where models need to be regarded "*as concrete artefacts that are built by specific representational means and are constrained by their design in such a way that they facilitate the study of certain scientific questions by means of construction and manipulation*" (Gelfert 2016, p. 113). At the same time, models can function as mediators in the mutual learning process between theory and data, as described in Box's feedback loop between theory and practice (Fig. 3.4).

If we use the standard characteristics for the statistical profession, such as 'statistics is the science of learning from data' or 'statistics means separating the signal from the noise', we need to interpret this in this iterative manner. In times of ubiquity of data for all facets of life, it is easy to forget this message, confusing data mining with the search for gold, as if valuable and interpretable information was hidden in

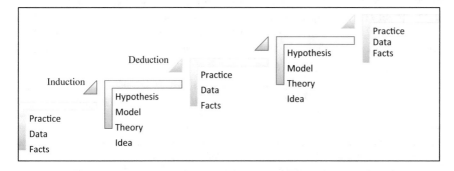

Fig. 3.4 Iteration between theory and empirical research. Adapted from Box (1976, p. 791)

the data-ocean.[21] If a comparison is to be made, then statistics assumes the role of a refinery, which uses the raw material 'data' to produce a veritable product.[22]

> No substantial part of the universe is so simple that it can be grasped and controlled without abstraction. Abstraction consists in replacing the part of the universe under consideration by a model of similar but simpler structure. Models, formal or intellectual on the one hand, or material on the other, are thus a central necessity of scientific procedure. ... That is, in a specific example, the best material model for a cat is another, or preferably the same cat. (Rosenblueth and Wiener 1945)

In social sciences, this relation between reality and a model of it has been introduced by Max Weber in form of what he called 'Idealtypen'. 'According to Weber's definition,

> an ideal type is formed by the one-sided accentuation of one or more points of view' according to which 'concrete individual phenomena ... are arranged into a unified analytical construct' (Gedankenbild); in its purely fictional nature, it is a methodological 'utopia [that] cannot be found empirically anywhere in reality' .... Keenly aware of its fictional nature, the ideal type never seeks to claim its validity in terms of a reproduction of or a correspondence with reality. Its validity can be ascertained only in terms of adequacy, which is too conveniently ignored by the proponents of positivism. (Kim 2012)

It is of fundamental importance for official statistics that conceptual models are designed in such a way that they are 'adequate' abstractions[23] of reality. This leads to the question of what 'adequate' concretely means or which criteria and which processes are offered by statistical methodology to answer this question. Compared to other quality components of statistical information (e.g. sampling errors), this area is less covered by statistical theory.

In German statistical terminology, 'Adäquation' (Grohmann 1985) represents the design phase within the process of statistical knowledge building, which contains basically the choice of model parameters according to the purpose of the research, available resources, time constraints, etc. "*Data quality is depending on ... developing operational methods corresponding as much as possible to theory and – intensively controlling and monitoring the survey procedure in process*" (Radermacher 1992).

This will say that statistical information is produced with two main ingredients: methodology and conventions. On the one hand, "*the notion of statistics as a primarily mathematical discipline really developed during the twentieth century, perhaps up to around 1970, during which period the foundations of modern statistical inference were laid*" (Hand 2009). On the other, the final products of statistical processes depend essentially on their conceptual design, which, as for other (manufactured)

---

[21] '*There is no such entity as GDP out there waiting to be measured by economists. It is an artificial construct ... an abstraction that adds everything from nails to toothbrushes, tractors, shoes, haircuts, management consultancy, street cleaning, yoga teaching, plates, bandages, books and all the millions of other services and products*' (Coyle 2015, p. 26).

[22] See the title of the Data Manifesto of the RSS '*What steam was to the nineteenth century, and oil has been to the 20th, data is to 21st*' (Royal Statistical Society 2014).

[23] It is important to keep in mind that adequate models nevertheless include (unavoidably) simplifications, since "*there is no accurate (or, rather, perfect) representation of the system that is simpler than the system itself*" (Cilliers 2000, p. 9).

products, depends essentially on whether the questions raised by stakeholders can be answered by statistics and whether they are answered in a satisfactory manner.

In order to be prepared for the following explanations with the necessary terms and their definitions, Desrosières is followed, who separates three *"aspects of statistics, (1) that of quantification properly speaking, the making of numbers, (2) that of the uses of numbers as variables, and finally, (3) the prospective inscription of variables in more complex constructions, models"* (Desrosières 2010, p. 114). Interestingly, the verb 'to quantify' is here defined and used differently from that of the verb to 'measure'.

> The idea of measurement, …, supposes implicitly that something real, already existent, analogous to the height of the Mont Blanc can be 'measured'… In contrast, the verb to quantify implies a translation, i.e. a transformative action, resulting from a series of inscriptions, codifications and calculations, leading to the making of numbers. This contributes to expressing and giving existence to, in a numerical form, something that before was only expressed by words and not by numbers." (Desrosières 2010, p. 115)

In addition, this representation is compatible with the information pyramid and the product portfolio of official statistics, as stated in Chap. 2. Finally, it is remarkable that Desrosières makes a distinction between 'numbers' and 'data'. Of course, his background is that of a survey statistician, where numbers are collected purposefully and earmarked and not, as with administrative or 'Big' Data, generated for other purposes and accessible from external sources.

After all these considerations, the question arises as to whether there are limits that official statistics cannot or should not exceed (Fig. 3.5). Here, it is important to distinguish between those limits that can be derived from an epistemological cause

Fig. 3.5 Limits of measurement

or limitation and those that are incompatible with the quality standards and public mandate of official statistics.

The first category includes, for example, hidden, not quantifiable phenomena or issues that lie in the future or those that belong to fiction. The example of Sustainable Development explains the issue: When it comes to achieving progress for the present generation without jeopardising the opportunities of future generations, it is expected that the relevant statistics will consider not only the present but to some extent the future. Whether and to what extent this should be a task for official statistics are the questions. Answering these questions positively, one accepts that this is a border shift that could have negative consequences for confidence in official statistics. Answering negatively, the official statistics deprive the great opportunity to become the responsible and central coordinator for the informational infrastructure of the strategically important policy area of Sustainable Development.

The second variety includes, for example, the monetary valuation of goods that are not traded on the market. The imputation of missing data (and also of prices) is part of the National Accounts. However, this is limited to values that can be derived from existing output data with clearly defined procedures. But when it comes to simulating entire non-existent markets and fictitious prices for goods (permissions for CO_2 emissions, for example), this presupposes heroic assumptions; this is certainly not part of the remit of official statistics.

Nevertheless, these limits of the second category are not immutable, but are partially pushed outward over time (e.g. from the programme of 2017 to 2025), so that previously excluded variables can be included in the standard programme. This shift is usually very carefully prepared, both in terms of development and in terms of the communication of the new results.

On the basis of a few examples, this is explained at this point:

- Well-being: Following the recommendations in the 2009 report of the Stiglitz/Sen/Fitoussi commission,[24] statistical measures of quality of life and subjective well-being were integrated for the first time in the 2013 EU-wide household survey EU-SILC (Eurostat 2015b).
- National Accounts: Subsequent to the revision of the UN standard SNA in 2008, the System of European Accounts ESA was revised in 2010 and successfully implemented in 2014 (Eurostat 2015a). It was prepared with immense effort including the communication, actually resulting in a smooth transition and great acceptance (as in the case of Quality of Life indicators). Nevertheless, in the media, the issue of assessing illegal activity[25] was focused on to a disproportionate degree.[26]

---

[24]The chapter '*Objective and subjective dimensions of well-being are both important*' and its recommendations deal with quality-of-life measures (Stiglitz et al. 2009, pp. 15–16).

[25]See, for example, http://europa.eu/rapid/press-release_MEMO-14-594_en.htm or https://www.istat.it/en/archive/115500.

[26]See, for example, https://euobserver.com/news/126110 or https://www.telegraph.co.uk/news/worldnews/europe/eu/11184605/Explainer-Why-must-Britain-pay-1.7bn-to-the-European-Union-and-can-we-stop-it-happening.html.

- Ecosystem Services Accounts: In 2012, a new UN statistical standard System of Environmental-Economic Accounting (SEEA) (United Nations 2014b) was adopted. This was a quantum leap in terms of the systematic quantification of ecology and economy under a methodological umbrella. However, this did not include the topic of ecosystem services[27] and in particular that of their economic, monetary valuation. The statistical community has so far not been able to develop consensus and widely supported methodology on this topic. While on the one hand, proponents argue that without such an assessment the important area of ecosystems is ultimately politically disadvantaged (because they are supposedly 'without value'), and opponents argue that such a serious value decision is not a technical matter (Kumar and Kumar 2008) and that 'fiction or simulation' does not fit official statistics when it is more than filling in marginal gaps in knowledge (Radermacher 2008).
- Experimental Statistics: The dilemma between preservation and renewal is difficult for official statistics to solve. Apart from the scarce resources and their allocation to recent statistics at the expense of existing ones, the general image of reliability and solidity is in danger, when suddenly products with a more innovative touch, but at the same time with more uncertainties and perhaps also errors, are contained in the programme. Escaping this trap is the aim of an 'experimental statistics'[28] project launched by Eurostat.

What is the lesson that can be learned from these examples? On the one hand, the area that is represented by official statistics is constantly evolving, so there is no fixed demarcation. On the other hand, there are topics that by their very nature do not lend themselves to quantification by official statistics, very often because they contain too many assumptions and value judgements and are not sufficiently based on observation. Thus, they do not meet the quality criteria of official statistics.

We will return to the components of evolution, adaptation and learning over time that are inherent in this presentation in the following sections.

3.2.3 Complexity

In the logical flow of this chapter from the questions of gaining knowledge with its possibilities and limitations to the questions of the application of knowledge in social processes, it is necessary to take a step aside to briefly address the subject of complexity. So, the next section will deal with questions that technically belong to mathematics about nonlinear dynamics. However, this is not about the mathematical-technical side of complexity, but about the conditions of gaining statistical information, the design and construction of metrics, and their possible analysis and synthesis (Maggino 2017).

---

[27] See https://seea.un.org/ecosystem-accounting.

[28] See http://ec.europa.eu/eurostat/web/experimental-statistics and (United Nations 2014c).

Of course, not all information needs and statistical observation objects are complex. Even if a topic is very broad, in many cases it is already possible to resort to a convention in which the producers and users of a statistic have agreed to reduce their very broad claim to some aspects that are easily measurable. In such cases, complexity has been reduced in advance and in the design process. However, this solution has dangers: in the interpretation, it must be emphasised that the measured sub-aspect is not necessarily representative of the broad whole. A good example of this is the GDP, which solidly measures one important aspect of the economy, namely the sum of all market transactions related with production, income and consumption. GDP was not meant for and is not designed to measure the complex phenomena of progress or of the welfare of a country. The reduction chosen in the design (for good reasons) of the indicator must be maintained in its interpretation. This example shows, however, that the temptation is great, yet to go further in the interpretation than permissible and somehow sell the measured part as a representative proxy to the whole (even if this relationship is unknown). How many times have you witnessed that the GDP was sold as a general welfare measure? The criticism of this indicator is unfortunately justified in such a false interpretation, but not in the calculation itself.

Thompson Klein (2004) emphasises that complexity in the subject matter might be coupled with the complexity of interdisciplinarity that is needed when dealing with such complex systems. She gives a number of examples for this difficulty to answer a question, solve a problem or address a topic that is too broad or complex to be dealt with adequately by a single discipline or profession, for example:

- Human interaction with natural systems (agriculture, industry, mega-cities);
- Major technical development (nuclear technology, biotechnology, genetics);
- Social, technical and economic developments interacting with elements of value and culture in ageing, energy, health care and nutrition;
- Sustainable Development.

That our modern world is 'complex' and that this complexity is rapidly increasing is one of the most common phrases today. Rightly, Peruzzi criticises the 'cult of complexity', which is based solely on a rhetorical use of the term, but does not have a scientific approach in mind (Peruzzi 2017, p. 3; Prigogine et al. 2017). The risk, then, is to use complexity as an excuse for difficulties in solving complicated issues, or simply as a mantra without sound scientific relinquishment. *"The current popular and scientific interest in the notion of complexity makes it one of the most prolific scientific research areas today. Although the idea of 'complexity theory' started as a 'scientific amalgam' emerging from the natural sciences, many of the concepts became popular through the appropriation and generalisation thereof in post-World War II developments in the fields of General Systems Theory, Cybernetics, and Artificial Intelligence"* (Woermann et al. 2018, p. 1).

Why is it, however, meaningful and necessary at this point to devote oneself to the topic of complexity when it comes to official statistics? It is the analysis of the relationship between knowledge and measurement,[29] which should prepare the ground for a closer look at science–policy interfaces[30] and the deduction of principles and general advice to official statistics concerning a strategic positioning in this political environment.

> An understanding of knowledge as constituted within a complex system of interaction would, on the one hand, deny that knowledge can be seen as atomized 'facts' that have objective meaning. Knowledge comes to be in a dynamic network of interactions, a network that does not have distinctive borders. On the other hand, this perspective would also deny that knowledge is something purely subjective, mainly because one cannot conceive of the subject as something prior to the 'network of knowledge,' but rather as something constituted within that network. The argument from complexity thus wants to move beyond the objectivist/subjectivist dichotomy. The dialectical relationship between knowledge and the system within which it is constituted has to be acknowledged. The two do not exist independently, thus making it impossible to first sort out the system (or context), and then identify the knowledge within the system. (Cilliers 2000, p. 8)

The complexity at issue here is a property of systems, both of the systems that are to be statistically quantified and the system that performs that quantification. A complex system is a system that consists of many components that can interact with each other. In these sort of systems *every element has systems above, below, and on the same level, related by any possible relationship: scale inclusion, control, etc. In consequence, there is no privileged perspective for an analysis of the whole system, which must be accepted as 'basic' or uniquely privileged. Nor there is a simple, linear causality among elements of the sort that is assumed in mainstream economics and in classical (but not in contemporary) physics. Accordingly, there is no possibility of conclusive knowledge of the total system'* (Ravetz 2018, p. 340).

The lessons that Julie Thompson Klein learned from an analysis of her above-mentioned examples are that:

- the research fields are ill-defined, a nexus of phenomena that are not reducible to a single dimension, their meaning is context dependent;
- there is no idealised model able to adequately and comprehensively mirror the behavioural pattern of the system;
- research is multilevel (micro, meso, macro);
- new forms of knowledge, institutional structure and problem solving require a new dialogue of science and humanities.

A statistician must, in the sense of 'profound knowledge', become familiar with the basic thoughts and the scientific aspects of complexity as far as they are relevant and necessary for professionalism in the respective field of application.

---

[29]The journal *Emergence* dedicated a special issue to 'Complexity and Knowledge Management' (Merali and Snowdon 2000).

[30]See van den Hove 'A rationale for science–policy interfaces' (Van den Hove 2007).

Box 3.1 Complexity

Main characteristics with relevance for statistics (Peruzzi 2017)

Data don't tell us how to interpret them, and when more than one interpretation is at hand, it is the theory we adopt which makes the difference (p. 3)

No property of an observed system S is independent of the observing system S' (p. 11)

Axes in research on complexity (p. 12):

- Co-evolution of a system and its environment
- Emergence of relatively stable systems through self-regulation
- Non-linearity
- Morphogenetic laws
- Sudden phase transitions
- Attractors of different shape for different systems

In social sciences, the subjective, or 'epistemic', side of complexity is no less important than the objective, or 'ontological', one. (p. 29)

This certainly does not mean that a statistician has to become an expert of the (mathematical) theory of complexity and relevant research. Nevertheless, a basic knowledge and an awareness of the importance of this topic is undoubtedly part of the building blocks of statistical training and statistical ethics (Stengers 2004) today. Without such basic knowledge and understanding, it is all too easy for complex issues to be disproportionately decomposed into their parts and quantified as supposedly independent of each other. The 'fallacy of misplaced concreteness' (Stengers et al. 2014; Daly 1987) is a risk that is only too prevalent in everyday statistical life.

A highly fragmented landscape, as we find in official statistics, is the result of long-term developments in which (by consensus between producers and users) a focus on isolated statistical 'silos' dominated. This resulted in a matrix of users and producers broken down by subject areas (e.g. agriculture, labour market, prices…), where essentially the main diagonal was of mutual interest This has changed radically, not least because the policy areas themselves are no longer so one-dimensional, so that for the purposes of agricultural policy, labour market statistics, prices, consumer data and environmental statistics are needed in addition to information about crop yields, livestock, etc.

Julie Thompson Klein has elaborated and generalised this observation:

Modern societies are increasingly ruled by the unwanted side effects of their differentiated subsystems, such as the economy, politics, law, media, and science. … postnormal[31] science is associated with 'unstructured' problems that are driven by complex cause-effect relationships. They exhibit a high divergence of values and factual knowledge in a context of intense political pressure. Hence, the stakes in decision-making are high, and epistemological and ethical dimensions are marked by uncertainties … complex problems are typically

---

[31] See *"Science for the post-normal age"* (Funtowicz and Ravetz 1993).

value-laden, open-ended, multidimensional, ambiguous, and unstable. Labelled 'wicked'[32] and 'messy', they resist being tamed, bounded, or managed by classical problem-solving approaches. As a result, the art of being a professional is becoming the art of managing complexity. (Thompson Klein 2004, p. 4)

No example could be better suited to portray these relationships, and the full range of complexities of both the statistical subject and the disciplines engaged than Sustainable Development. We will discuss Sustainable Development at the end of this chapter in Sect. 3.5. Before that, however, we still need some terms and tools from the social sciences that will help us to better understand the place and role metrics can and should have in the societal and political process for achieving the UN's strategic goals in 2030.

3.3 Statistics and Society

The relationship between (official) statistics and society basically has different forms, which are worth analysing one after the other:

- First, questions of **epistemology** arise. Is the process of statistical collection of data and the generation of information itself part of a social process, so that it cannot be assumed that this cognitive formation is independent?
- Secondly, the scientific approaches of **sociology** are needed, with which (and with their tools, terminology, etc.) such interactions are analysed.
- Third, it is crucial to consult the scientific work-up on the **history** of official statistics. History does not repeat itself, at least not in the same way. Nevertheless, looking to the future can benefit from the understanding of previous periods.[33]
- Finally, it is a question of using prudent **governance (including ethical questions)** to ensure that the benefits inherent in official statistics are not undermined, endangered, or reversed by naivety or cynicism, excessive expectations, or (negligent or intentional) abuses.

In the following, first the tools developed in sociology for the more general application of interactions between Science, Technology and Society (Science and Technology Studies, S&TS[34]) will be introduced as far as this is relevant for an outlook on Official Statistics 4.0.

---

[32] See also "Technology Run Amok - Crisis Management in the Digital Age" (Mitroff 2019).

[33] *"Zukunft braucht Herkunft"* (the future needs the past) (Marquard 2003). *"Das Neue, das wir suchen, braucht das Alte, sonst können wir das Neue auch gar nicht als solches erkennen. Ohne das Alte können wir das Neue nicht ertragen, heute schon gar nicht, weil wir in einer wandlungsbeschleunigten Welt leben"* Interview with Odo Marquard in *Der Spiegel* 9/2003 (http://www.spiegel.de/spiegel/print/d-26448590.html).

[34] An introduction to S&TS can be found in https://en.wikipedia.org/wiki/Science,_technology_and_society; overviews provide (Restivo 2005) or http://stswiki.org/index.php?title=Worldwide_directory_of_STS_programs or for example the UCL Department of Science and Technology Studies http://www.ucl.ac.uk/sts.

3.3.1   Co-construction, Boundary Object, Governance

If a statistician deals with the relevant literature and research in the social sciences, then it is necessary to engage in a different way of thinking and working. Definitions, concepts and tools initially appear imprecise, poorly defined and their purpose seems vague. One misses the unambiguity that is, for example, in mathematical formulas or quantitative evidence. There is a clash between the mathematical mind and the sociological mind, which represents a permanent challenge for official statistics.

This also applies, in particular, to the notions to be presented now—'co-production', 'boundary object' and 'governance'[35]:

Co-construction (or co-production, mutual construction, co-evolution) highlights the mutual influence between producers and users of a technology, such as statistics, and elaborates *"the simultaneous processes through which modern societies form their epistemic and normative understandings of the world. This framework, most systematically laid out in 'States of Knowledge' (Jasanoff 2004b), shows how scientific ideas and beliefs, and (often) associated technological artefacts, evolve together with the representations, identities, discourses,* and *institutions that give practical effect and meaning to ideas and objects"*.[36] Data, information and knowledge are human products of social processes; they are influenced by the way we see the world and they are mutually influencing the world we live in (Saetnan et al. 2012). Co-production studies therefore seek to explore how *"knowledge-making is incorporated into practices of state-making, or of governance more broadly, and in reverse, how practices of governance influence the making and use of knowledge"* (Jasanoff 2004a, p. 3). Already, the statistical activity of classifying is partly of far-reaching influence on our behaviour and daily life.[37]

Boundary object   is, although widely unknown, a fruitful concept waiting to be applied in statistics. Generally, those objects can be things (such as statistical information) or theories that can be shared between different communities, with each holding its own understanding of the representation. They are *"both plastic enough to adapt to local needs and constraints of several parties employing them, yet robust enough to maintain a common identity across sites … They have different meanings in different social worlds but their structure is common enough to more than one world to make them recognizable, a means of translation"* (Star and Griesemer 1989, p. 393)[38].

---

[35]This is also emphasised by Sheila Jasanoff: *"However, co-production, in the view of contributors to this volume, should not be advanced as a fully fledged theory, claiming lawlike consistency and predictive power. It is far more an idiom – a way of interpreting and accounting for complex phenomena so as to avoid the strategic deletions and omissions of most other approaches in the social sciences"* (Jasanoff 2004b, p. 3).

[36]See Sheila Jasanoff's introduction on her website https://sheilajasanoff.org/research/co-production/.

[37]One of the pioneers of this field, Susan Leigh Star, has demonstrated this in 'Sorting Things Out' using a variety of examples (Bowker and Star 2000).

[38]Quote in (Saetnan et al. 2012, p. 9).

Although similar to 'instruments', 'boundary objects' are different in their function: While boundary objects are two-sided (belonging to two or more disciplines), instruments are one-sided (Stamhuis 2008, p. 364). The difference can be explained simply by drawing a parallel to languages. Languages with their syntax, grammar and semantics fulfil their functions within given framework conditions and contexts. Like a regional dialect, a professional jargon is geared to the communicative needs and abilities of a specific community. A completely different situation exists when international discussions take place (be it at conferences or international projects) where participants of different native languages use a common working language. Although the language is formally in all cases, e.g. English, there will be immense differences between these forms. The English used in the international context comes close to a 'boundary object'.

In official statistics, this distinction becomes important from where the results leave the sphere of statistics and are used by different users for their purposes and needs. From this point on, it is no longer enough to speak the language of statisticians. Rather, it requires a form of communication that is also understood 'on the other side'. Especially in the conception of indicators, it must be assumed that they are 'boundary objects' in their function. This changes a lot for those who work on their design. It is not enough in this context to develop such indicators from a purely statistical point of view. Rather, they must be optimised from the outset to their later purpose and area of application, which puts the communicative aspects in the centre.

The term '**governance**' is one of those that combine two qualities: firstly, they are poorly defined and, secondly, they have settled in our language at great speed. That may seem paradoxical. Maybe these two properties fertilise and reinforce each other. In addition, it is difficult to find a general technical term that is also open to linking legal regulations and voluntary commitments, ethical principles and other forms of policies or political commitments.

'Governance', 'government' and 'governmentality' are closely related but not identical notions that unfold different concepts and approaches. In his analysis, Foucault (1991) pushes the furthest into various forms of governance, not only at the level of entire states, but also on a small scale, that is, in the management of companies, hospitals, etc. Foucault describes in detail the historical development towards what we take for granted today as a common form of governance and management. From the description of different forms, it is then only a small step to the presentation of guidelines for 'good governance', thus a rather normative dimension.

Box 3.2 Data for policy and return

The growing importance of data and information for political decisions is reflected in the handy and popular formulation 'Data for Policy' (D4P). However, this label is only suitable to a limited extent for characterising the network of relationships between data and politics.

Even if the amount of data is growing at an enormous rate in the era of digitalisation and globalisation, this raw material of data is not directly usable for politics. Instead, suitable processes are needed to distil, refine and process the valuable content for politics into digestible information from the flood of raw data. The term 'facts' is used here as a generic term for such information. When data is at the beginning of the processing, facts are at the end.

With this distinction between data and facts, it is now possible to break down the relationship between policy and data into different components.

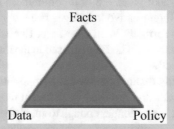

The following relationships must be distinguished from each other:

Data to Facts (D2F): This is the field for statisticians, data scientists and empirically working researchers from different disciplines; facts can mean results of standard statistical processes on the one hand, but on the other hand also, for example, unique research-based evaluations of micro-data.

Facts to Policy (F2P): This is the field of work of the specialists who prepare and use the information content of the facts for policy advice; journalists and researchers with future-oriented models are among them.

Policy to Facts (P2F): On the one hand, this is about the policy-relevant design of facts (indicators, accounts, indices, maps, graphs, etc.); on the other hand, it also includes questions of governance (who participates in the design process, who ultimately decides on the selection of the statistical programme, how much money and time is available, etc.).

Facts to Data (F2D): This is the scientific and technical conception of the generation or selection of suitable data sources; questions of authorisation, confidentiality, accessibility and ownership of the data are included.

Policy to Data (P2D): In many ways, politics sets the framework conditions for the generation of data, for their protection, for the infrastructure of research institutions or data centres, for the design of legal conditions, such as copyrights; politics also influences the economic framework conditions under which industries develop innovatively and competitively in the digital era.

Data to policy/politics (D2P): New data and inno- vative methods of data-science can be used for exper- imental statistics which can be helpful in the policy—making process to guide priorities and identify future policy issues.

Fig. 3.6 Co-construction of information and institutions. Graphical adaptation of Soma et al. (2016, p. 132)

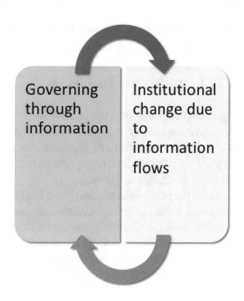

In principle, however, it is important to see that governance has two sides, namely firstly (active) governing through information and secondly governance as institutional framework. Information is very important in any form of governance: knowledge is power. Moreover, a feedback loop between information and social institutions in the sense of co-production is a fundamental part of any form of governance: on the one side, the governance through information with the purpose to guide, steer, control or manage sectors or facets of societies; on the other side, changes in institutions caused by the massive increase of information processes (Fig. 3.6).

At this point, however, it is crucial to dispel the quantum leap made by information as an instrument of governance in recent historical development. In this development, information enters the centre of power, becomes an instrument of explicit or covert influence, serves as an attack and counterattack and finally presents itself absurdly in the battle for facts against alternative facts.

"*The Politics of Large Numbers*" (Desrosières 1998) is the title of Alain Desrosières famous book. Like Desrosières, Hacking (1991) and Porter (1995) have also analysed and described the historical co-evolution of governance and statistical information.

The history of these interactions is multifaceted and anything but linear. The following section takes a very brief note of the essential aspects in so far as they seem helpful in understanding the current situation.

3.3.2   The Co-construction of Statistics and Society—History in Fast Motion

3.3.2.1   200 Years of Statistics, Science, Society

The term 'statistics' has the same linguistic roots as 'state'. In every form of governance, rulers needed statistical information, information that allowed them to target policy. The current form of statistics, however, has experienced its first great blossoming with the birth of the nation state and has since developed along with this in all its variants, ups and downs.

Over the past two centuries, there has been a constant interplay of impulses from statistical-methodological innovations, new data and data processing and socio-political framework conditions and influencing factors. This evolution has been largely continuous but has also been marked by major upheavals and changes. Some of these characteristic historical moments and periods are of particular value in understanding our present and nearer future.[39]

3.3.2.2   The Making of the 'State'—National or European

There is hardly a more instructive example of the importance of statistics for the creation of the nation state than the political developments in Italy in the first half of the nineteenth century. As analysed by Silvana Patriarca in her book *Numbers and the Nationhood* (Patriarca 1996) ('patriotic'), statistics in and by the so-called Risorgimento were used to define the identity of the 'Italians' and to draw up a political map of the Italian nation. So, it happened that there was a statistical yearbook for Italy before this nation emerged from a union of the different territorial parts and became politically institutionalised.[40] Many of the questions that still concern us today were discussed extensively and controversially in this epoch as well as in the period after the founding of the state. This includes, in particular, the current question of the demarcation of statistical versus political facts or the importance of positivism for statistics.[41]

Similarly, the history of official statistics at international level can be linked to the corresponding historical phases on the political side (e.g. United Nations, OECD). Of course, these political institutions cannot be directly compared with a nation state; the governance and political mandate of an international body are completely different. Nonetheless, these institutions also need good-quality statistical information for their work. The establishment of a political structure is therefore necessarily accompanied

---

[39] A historical understanding of macro-economic statistics is provided in Daniel Mügge's *"The Revenge of the Political Arithmetick. Economic Statistics and Political Purpose"* (Mügge 2019).

[40] *"The Annali universali di statistica signaled this new phase by starting to publish in 1852 a column called Cronica statistica Italiana"* (Patriarca 1996, p. 148).

[41] *"Nor did Italian practitioners abandon the idea that statistics was essentially a governmental science and had eminently civil function"* (Patriarca 1996, p. 186).

by a statistical infrastructure whose governance reflects the relationship between the international level and national actors.

This becomes very clear for statistics in Europe. Already at the beginning and during the first years of the European Coal and Steel Community,[42] the European Union had set up a public statistical service. Out of this small service of international cooperation, over six decades, in several relapses and along with the European treaties, the European statistical system has emerged, a system that corresponds to the supranational character of the European Union today (De Michelis and Chantraine 2003).

It is therefore to be expected that the future of official statistics will continue to be heavily influenced by (and vice versa will have an influence on) forms of governance that will develop nationally and internationally (or perhaps regionally). At the moment, it is probably harder than ever to foresee this future. Will the forces assert themselves, which demands a re-strengthening of the nation state, or will globalisation bring about its weakening (Dasgupta 2018)? Will the European Union be reformed and possibly strengthened? There are currently no adequate answers to such questions. One thing is certain, however, in the development of any variant of the state, statistical information is needed to a large extent, as can be seen at the moment in the Brexit negotiations. In the end, as in the Italian Risorgimento of the nineteenth century, there is the question of a 'European people' and how this can be statistically defined and promoted.[43]

3.3.2.3 Ideologies and Their Influence

An interesting feature in the history of statistics has to do with statistics and statisticians being children of their own time. What is important or unimportant, which questions are pursued scientifically or empirically, and which mental attitude or conviction play a role in the work of a statistician, is generally not decided objectively, but bears the stamp of the historical episode. In the opposite direction, eminent scientists of statistics have also influenced the political events of their time.

A first example, which could highlight the mutual influence between ideologies and statisticians, is the case of 'eugenics':

> The term 'eugenics' was coined by Sir Francis Galton in his 1883 book "*Inquiries into Human Faculty and its Development*" and derives from the Greek 'eu-genes', meaning 'well-born.' There Galton defines eugenics as 'the science of improving stock – not only by judicious mating, but whatever tends to give the more suitable races or strains of blood a better chance of prevailing over the less suitable than they otherwise would have had.... eugenics aims to use science for human improvement over generations by changing the composition of human populations through favouring the reproduction of certain sorts or kinds of people. Although Galton characterized eugenics as itself a science, it was also a social movement, one that gained traction in many countries early in the 20th-century.[44]

---

[42]See, e.g., https://en.wikipedia.org/wiki/European_Coal_and_Steel_Community.

[43]See the research project 'Arithmus: peopling Europe: how data make a people' (Goldsmiths 2018).

[44]Robert A. Wilson in Eugenics Archives (Cassata 2017).

In 1907, Karl Pearson became the first director of the Galton Laboratory for National Eugenics, established at University College London (UCL).

> In Italy, *"anxieties over national regeneration, technocratic ambitions and new social welfare-oriented policies, which, after the war, accompanied the crisis of the last liberal governments and the progressive rise of fascism, favoured the affirmation of eugenics as a part of social medicine and public health. In this context, eugenics was progressively seen as a paradigm of national efficiency, based on the subordination of individual liberty to superior collective interests for the 'defence of society and the race.' leadership mirrored this ideological and political fusion: the president was the demographer and statistician Corrado Gini..."* (Cassata 2017).

The Eugenics Archive of Canada[45] allows an interactive study of the relationship between people, events and concepts in this important area, which is not very present nowadays (Fig. 3.7).

Starting with Adolphe Quetelet (l'homme moyen) (Quetelet 1835; Desrosières 2002), through Francis Galton and Karl Pearson (Porter 2004) to Corrado Gini (Quine 1990; Horn 1994), the undoubtedly creative intellectual elite of statistics in the second half of the nineteenth and the first decades of the twentieth century was rooted to an extent in an ideology that seems strange from today's perspective.

A second instructive example of the interaction between ideologies, method development and statistical leaders is provided by the career of Rolf Wagenführ, who started in the 1920s his scientific way at Jena University with a thesis on Soviet

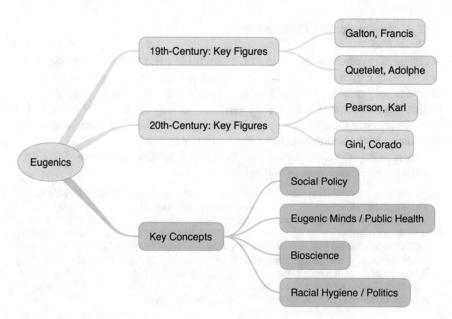

Fig. 3.7   Eugenics. Extract from interactive source Eugenicsarchive (2018)

---

[45] See http://eugenicsarchive.ca/.

business-cycle theory and in 1933 with a study of Germany's long-run industrial development (referring to Marx), to then turn to publications on issues of rearmament and military economics, being appointed as chief industrial statistician of the Institut für Konjunkturforschung in 1943 and further chief statistician in Speer's Ministry for Armaments, in charge of the statistical information system for running the war economy in 1945. After the war, he remained a wanted man, first working for the central command of the Soviet military administration of Germany, then 'transferring' to the US zone, where he worked for John Kenneth Galbraith's Overall Economic Effects Division of the United States Strategic Bombing Survey, which assessed the economic effects of allied bombing during the war. In the late 1940s, he was chief statistician to the West German trade union federation before finally, in 1952, being placed in charge of the temporary statistical department of the European Coal and Steel Community (ECSC), later first Director of Statistics for the European Community (EC) until 1966 (Tooze 2001; Fremdling 2016).

Although (or perhaps because) Rolf Wagenführ's biography was marked with many dramatic twists, it may not be untypical of life histories of the time: influenced by ideologies and political currents and in turn influencing them.

The influence of ideologies on statistics, on method development and on statisticians has been reflected in a work of art by artist Arnold Dreyblatt, which can be seen in front of today's 'Center for Studies of Holocaust and Religious Minorities' in Oslo, Norway. The Centre has installed "*Innocent Questions, an artwork by Arnold Dreyblatt in front of the building, which During World War II, was the home of the leader of Norway's pro-Nazi puppet government*" (Minorities 2018). Nothing could better raise the question of the guilt or innocence of statistics[46] than this simple sculpture of a punch card of population statistics (Dreyblatt and Blume 2006).

The demographer William Seltzer contributed to the conception of this art object. He concludes:

> *...statistics are important for governments, policy makers, and the people themselves. ... Such data help in decision making, ... and policy-oriented research. ...Taken by themselves and in most circumstances, the collection of such information ...involves 'innocent questions'. Unfortunately, this is not always the case. Particularly when information is also obtained that permits the identification of vulnerable population sub-groups through questions on race, religion, ethnicity ... the results have sometimes been used by State or those in power to target vulnerable population subgroups for adverse action, including human rights abuses. ... The point here is not to discourage the collection and use of population statistics. Rather, it is to remind us all ... that we all carry a heavy obligation to ensure that our national statistical systems are not diverted from their legitimate purposes to the kind of misuse portrayed here. We all have a continuing responsibility to guard against such misuses* (Seltzer 2006, pp. 71–75) (Fig. 3.8).

---

[46]See also (Wietog 2003).

Fig. 3.8   Innocent Questions, Arnold Dreyblatt. Photos courtesy of the artist Arnold Dreyblatt

3.3.2.4   The Audit Society, Neo-liberalism and Populism

'*The ability to take into account the consequences of a decision, together with their unintended feedbacks, was always recognised as an ingredient of rationality, of direct relevance to any strategy in politics, business and conflict*' (Peruzzi 2017, p. 4). According to this statement, one behaves rationally if one knows the advantages and disadvantages of decisions and decides soberly on this basis. In what form and with what means these advantages and disadvantages are concretised, and with which criterion finally a decision is made, is still open.

'*If you can't measure it, you can't manage it!*'[47] is praised today as a guiding principle in modern management, interpreted as a consistent and up-to-date implementation of the rationality principle. In times of the data revolution, when data about everything is always available, why should it not make sense to make decisions based on hard facts, based on evidence?

Apparently, decision-making underwent a massive '*quantitative turn*' (Sangolt 2010a, p. 75) during the last century, driven by the availability of data and (statistical) methods to compute information, driven also by decision-making algorithms and standards in Economics, Management, Audit or Operations Research, and last but not least by the contemporary form of governance. Michael Power in his book the "*Audit Society – Rituals of Verification*" has already in the 1990s expressed his view, "*that we are in the middle of a huge and unavoidable social experiment which is conspicuously cross-sectional and transnational*" (Power 1994, p. 16).

---

[47]For an introduction and critical review, see https://blog.deming.org/2015/08/myth-if-you-cant-measure-it-you-cant-manage-it/.

The following section reflects on the significance of an 'evidence-based decision-making' or, more recently, a 'data-driven decision-making' environment for official statistics. The analogy to a medicine is used, the effect of which crucially depends on the fact that a medicine is applied with the right dose and the right timing. Deviations from the prescribed dosage may cause side effects, short-term and long-term damage; permanent misuse may lead to dependencies.

The benefits generated by information-based, rational decision-making are immense. Instead of vague opinions and incomprehensible subjective impressions, objectively measured and quantified facts occur and are an important element in the knowledge base. This shortens the decision-making process, prevents or reduces conflicts, increases transparency both internally and externally, and allows for controls in the implementation of the decision. More targeted actions are possible, improving their effectiveness and efficiency. Benchmarking opens the door to an exchange of good practices and mutual learning. Management by objectives enables a modern and emancipated approach between superiors and employees. Quality improvements are realised through constant measurement and learning.[48]

As far as decision-making in the public sector is concerned, the claim is formulated in the so-called Data Manifesto of the Royal Statistical Society (Royal Statistical Society 2014): *"Evidence must be taken more seriously in policy formulation and evaluation, and statistics should be at the heart of the policy debate. Making policy when resources are tight is difficult but choices should take into account the probable quantified consequences of alternatives. … Government should publish the data and evidence that underpin any new policy it announces and should also commit to regular and long term evaluation policies. Where we lack the data to inform choices between options in important policy areas, we should invest in getting it"*.

That better data means better decisions is not a new discovery. Nevertheless, if this motto is used by the UK Statistical Authority,[49] it is an indication that, surprisingly, it seems necessary to promote statistics—apparently a paradox. There seems to be a contradiction between the principle that evidence is essential and the interest that this evidence is and can be produced with good quality.

A snapshot from a practitioner's blog gives an impression of what the problem is and how the contradiction can be explained: *"The good news is that we manage these unmeasurables perfectly well without any need for yardsticks. 'If you can't measure it, you can't manage it' won its place in the Big Book of Business Dogma because the business world, or at least the bureaucratic edifice it relies on, the one we call Godzilla, is all about measurement. Measurement is a religion in the business world! If we can slap a metric on something, by God, we're going to do it. We love to measure things, because it makes us feel as though we're really doing something. … Measurement is our drug in the business world…"* (Ryan 2014).

---

[48]One of the currently leading quality management approaches Six Sigma defines the concept as follows: '*Six Sigma is a systematic approach to process improvement using analytical and statistical methods. The special feature of Six Sigma compared to other process improvement methods is the mathematical approach. It is assumed that every business process can be described as a mathematical function*' http://www.six-sigma.de/en/six-sigma-definition.

[49]https://www.statisticsauthority.gov.uk/better-statistics-three-years-on/.

It is worth remembering the theorems of W. E. Deming, which were explained at the beginning of this chapter. Deming has warned of a measuring machinery. Excessive measurements jeopardise the positive effects of the medicine 'evidence' and create (of course unwanted) side effects that are ultimately detrimental to the goal of good data and statistics. Appreciation of information, awareness of limits of measurability and unconditional prioritisation of the quality of statistics are key factors. If these factors are not sufficiently strong, one is only guided by the presence of data; quantity goes before quality.

However, there is a larger context that is important in understanding the current status of statistics in Western societies. Ever since Michel Foucault dealt with the forms of governing, the relationship between knowledge, power and techniques of governance has been explored by scholars of different disciplines. Michel Foucault: *"The theory of the art of government was linked, …to the whole development of the administrative apparatus of the territorial monarchies, the emergence of governmental apparatuses; it was also connected to a set of analyses and forms of knowledge …which were essentially to do with knowledge of the state, in all its different elements, dimensions and factors of power, questions which were termed precisely 'statistics', meaning the science of the state…"* (Foucault 1991, p. 96).

Further, Foucault relates 'economy' with 'governance' and with 'statistics' in the following manner: *"It was through the development of the science of government that the notion of economy came to be re-centred on to that different plane of reality which we characterize today as the 'economic', and it was also through this science that it became possible to identify problems specific to the population; but conversely we can say as well that it was thanks to the perception of the specific problems of the population, and thanks to the isolation of that area of reality that we call the economy, that the problem of government finally came to be thought, reflected and calculated outside of the juridical framework of sovereignty. And that 'statistics' … now becomes the major technical factor, or one of the major technical factors, of this new technology"* (Foucault 1991, p. 99).

Finally, Foucault introduced the famous term 'Governmentality', which since then has become a scientific term that inspired an entire branch of sociological studies and research.[50] 'La Gouvernementalité',[51] in his view, means three things:

1. The ensemble formed by the institutions, procedures, analyses and reflections, the calculations and tactics that allow the exercise of this very specific albeit complex form of power, which has as its target population, as its principal form of knowledge political economy, and as its essential technical means apparatuses of security.

2. The tendency which, … has steadily led towards the pre-eminence over all other forms (sovereignty, discipline, etc.) of this type of power which may be termed government, resulting, on the one hand, in the formation of a whole series of specific governmental apparatuses, and, on the other, in the development of a whole complex of saviors.

---

[50] A small sample of references might be sufficient here (Bröckling et al. 2000; Burchell et al. 1991; Ewald 1991; Hacking 1991; Zamora and Behrent 2014; Fried 2014; Hammer 2011; Jasanoff 2004b; Brown 2015; Sangolt 2010b; Lupton 2013; Davies 2016; Power 1997).

[51] Foucault (1978).

3. The process, or rather the result of the process, through which the state of justice of the Middle Ages, transformed into the administrative state during the fifteenth and sixteenth centuries, gradually becomes 'governmentalized'." (Foucault 1991, pp. 102–103)

The so-called neo-liberal form of governance that Foucault describes is controversial in many aspects, particularly in terms of the penetration of life through economic principles and the focus on economic goals such as efficiency and competition. However, this is not of direct relevance to this work. Rather, the point here is to show the quantitative turn that results from the fact that in modern Western societies, it is a supposed normality to do a "*systematic effort to delineate and measure the objects and results of governance quantitatively for the purpose of demonstrating competitive edge and superiority at the individual and/or collective level*" (Sangolt 2010a, p. 75).

When Linda Sangolt speaks of the '*Age of Quantification and Cold Calculation*', she refers to the twentieth century and its waves, fashions and policies (such as Taylorism and Thatcherism), which have contributed to accelerate and intensify the use of metrics for governance purposes in the private and the public sectors (Sangolt 2010a, p. 77). In particular, its impact on the public sector, for example in 'New Public Management',[52] is of interest to this work. As mentioned earlier, it cannot be ruled out that the same policy calls for data, but questions its production. The example of Sir Derek Rayner's activities in Margaret Thatcher's public sector reform[53] shows that statistical offices are among the first targets of austerity policies, despite the fact that they provide much needed political data. It can be concluded that statistics have power, but statisticians do not.

Desrosières describes the history of statistics[54] in relation to that of economic theories and that of economic policies in a very condensed overview of crucial phases and forms of governance as well as their impact on statistics (see Fig. 3.9). The history of statistical methods is linked "*with the history of issues placed on the agenda for official decisions which themselves subsume: (1) ways of conceptualizing society and the economy, (2) modes of public action, and (3) different forms of statistics and of their treatment*" (Desrosières 2011). From this overview of the historical stages of development, one can also deduce how the current official statistics programme has developed over the different forms of governance.

According to this overview, we are currently in the phase of neo-liberal governance, which seeks to achieve self-regulation of actuaries using measures and indicators rather than instructing them with regulations. This approach promises no less

---

[52]For a critical review see (Lægreid and Christensen 2007).

[53]For the review of the UK Statistical Service under M. Thatcher see GreatBritain (1981), Thomas (1984).

[54]"*Statistics is not only, as a branch of mathematics, a tool of proof, but is also a tool of governance, ordering and coordinating many social activities and serving as a guide for public action. As a general rule, the two aspects are handled by people of different specializations, whose backgrounds and interests are far apart. Thus, mathematicians develop formalisms based on probability theory and on inferential statistics, while the political scientist and sociologist are interested in the applications of statistics for public action, and there are some who speak of 'Governing by indicators'. The two areas of interest are rarely dealt with jointly*" (Desrosières 2011, p. 41).

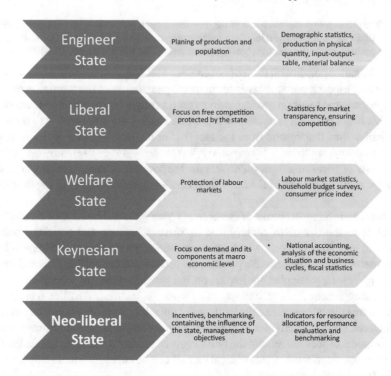

Fig. 3.9 The state, the market and statistics. *Source* Desrosières (2011, p. 45)

than modernity, democratisation and transparency. Furthermore, flattening of hierarchies and auditability[55] are to be ensured as well as a general quality improvement. Improving efficiency is the overall goal of this approach.

For a long time now, this form has become a good governance standard, not only in the private but also in the public sector. The 'competitive strategy' that Michael Porter had presented in 1980 for the management of corporations (Porter 1980) was transferred as advice for the achievement of an 'competitive advantage' for the policy-making of entire states in 1990 (Porter 1990). Under the regime of New Public Management (Lægreid 2017), no public initiative could possibly be launched and justified without a cost-benefit analysis or an impact assessment. Hospitals and universities operate as well under the control of key performance indicators as municipal administrations.

The intensity and speed with which this form of governance has penetrated into all areas of social life depends on cultural circumstances and political initiatives. A review of official statistics marked, for example, the beginning of the efficiency-oriented reforms of the Thatcher government in the early 1980s. As already mentioned, the inconsistency in this policy is still astounding: on the one hand, data

---

[55]Michael Power sees even *"audit as intrinsic to modern society – 'a constitutive principle of social organizations' and an 'institutional norm'"* (Power 1994).

and information are needed for 'modern' governance, as never before; and on the other, the Statistical Office is envisaged as the first target for cuts in the supposedly costly public administration. Since these beginnings, especially in the Anglo-Saxon countries, a 'stealth revolution' (Brown 2015) has taken place. The most far-reaching variant is not just about quantification but even a monetary assessment of each and every aspect of life (Cook 2017).[56]

Nevertheless, one may ask what could be wrong about it if transparency and efficiency are the lifeblood of modern management. To answer this question, one has to deal with the side effects of this 'medicine'.

The side effects can be particularly significant if the information that plays a crucial role is not treated properly. This improper treatment can take place in different forms. For example, it is fatal for the requirement to quantify desired target values (e.g. key performance indicators) to disregard any limitations and restrictions for obtaining high-quality information.[57]

Evidence-based decision-making enormously reinforces the importance of statistics. However, statisticians have to be more aware of their responsibilities and not only familiarise themselves with the demands for information but also with the associated risks.

An extensive literature provides information on the most varied forms of the interaction between governance and statistical quantification, with possible consequences on both sides, statistics and decision-making.

A very serious risk for statistical quality is recorded in the so-called 'Goodhart's Law': "*Any observed statistical regularity will tend to collapse once pressure is placed upon it for control purposes*".[58] As soon as an indicator is to be taken seriously, because consequences are linked to its amount, it is tempting to manipulate the measurand. If you do not like the weight, which indicates the bathroom scale, you can adjust the scale. This turn could then be called 'decision-based evidence-making'.

However, this feedback loop between measurement and behaviour is by far not the only one.

Another well-known risk is the one called 'streetlight effect': "*The first step is to measure whatever can be easily measured. This is OK as far as it goes. The second step is to disregard that which can't be easily measured or give it an arbitrary quantitative value. This is artificial and misleading. The third step is to presume that what can't be measured easily really isn't important. This is blindness. The fourth step is to say that what can't be easily measured really doesn't exist. This is suicide.*"[59]

---

[56]See, for example, "The Club of Rome to William Nordhaus and the Nobel Committee: 'Pursue Profitability – Even at the Cost of the Planet?!'" (Dixson-Declève et al. 2018).

[57]Overpowered metrics eat underspecified goals (https://www.ribbonfarm.com/2016/09/29/soft-bias-of-underspecified-goals/).

[58]Quote from https://en.wikipedia.org/wiki/Goodhart%27s_law.

[59]Yankelovich, D., in *Corporate Priorities: A Continuing Study of the New Demands on Business*, cited in http://pdf.wri.org/bell/tn_1-56973-179-9_teaching_note_english.pdf.

This streetlight effect comes into play, for example, if the question is to be answered whether the GDP is a measure of the welfare or the progress of a country. Undoubtedly, the GDP is neither constructed nor suitable for such very broad statements. Nevertheless, in the general opinion formation such overvaluation and misinterpretation take place, which in turn give rise to a GDP criticism, which actually lacks any foundation.

General threats to statistics are associated with a misunderstood measurement 'culture'. When the 'streetlight effect' occurs, that is, negating that there is a gap between what one wishes to measure and what one can actually measure, then this will cause the gap in one way or another to be filled, whether by unsound estimates, by an inappropriately high effort or simply by fake facts. One aspect of a false culture of measurement has already begun to appear: the high expectation pressure with regard to quantification can lead to exhausting the desired activities in building up a measuring machinery, which could then lead to evidence only, without any use in form of decision-making.

Finally, there are long-term implications for the way in which decisions and, in particular, political decisions are prepared, how they are communicated and implemented, and, above all, how they are perceived by various stakeholders, which are the citizens in political decisions. If indicators are used primarily to shorten the public discussion ('closing-up'), this can be perceived as the dominance of technocrats, as inadequate participation and transparency, as a form of 'evidence-instead-of-decision-making'.[60]

For a few years, however, there has been growing resistance to the widening of data-based governance (as part of a liberal, Western type form of modernity), which unfortunately is driven by the advocates of populism. The Oxford Dictionaries and the Society for the German Language have both chosen "*post-truth*"[61] (or in German "*post-faktisch*") (GfdS 2016) as Word of the Year 2016. Growing scepticism about all forms of experts does not stop at journalists or at statisticians. Coupled with a lack of statistical literacy and the impression of being at the mercy of the representatives of a perceived technocratic regime, the resistance is formed,[62] which not only reduces the side effects of the medication 'statistics', but—comparable to the resistance to general vaccination—argues from a gut feeling, that completely negates the existence of facts and thus throws the benefits and progress of enlightenment overboard (Davies 2017; Pullinger 2017).

So, when talking about the 'Data Revolution', the second revolution, that of the neo-liberal governance, and the third, the populist resistance to expert opinion, should

---

[60] An example is the Journal Impact Factor and its "*enormous effects on the scientific ecosystem: transforming the publishing industry, shaping hiring practices and the allocation of resources, and, as a result, reorienting the research activities and dissemination practices of scholars*" (Larivière and Sugimoto 2018, forthcoming, p. 2).

[61] See https://www.oxforddictionaries.com/press/news/2016/12/11/WOTY-16.

[62] "*Wenn Technokraten über politische Fragen entscheiden sowie politische Beschlüsse fassen und die Sparpolitik sowieso schon gefährdete Gruppen hart trifft, schafft das Nährboden für die Abweisung und Politisierung von Expertenwissen auf breiter Front*" (Hendricks and Vestergaard 2018, p. 128).

all be mentioned in the same breath.[63] All three developments are related to each other and from each other in a certain way. All three together must be perceived as a social framework for official statistics in the present and in particular in the future.

Under the influence of the dynamics generated by machine-to-machine communication, artificial intelligence (AI) and algorithms, these observed trends will gain speed and relevance for each and every segment of our life.[64] The phrase 'evidence-based decision-making' was used to express that decisions could and should be augmented by the use of information. What if they are not only augmented but automated? Can it be sufficient to screen the 'horizons for a data-driven economy'[65] focusing only on the technical aspects of this major change process? The answer to these questions, of course, is that the pros and cons, the effects and side effects, will continue to increase and accelerate. Aspects of ethics and governance as well as those of communication and participation, and of education and data culture, will become more important. It is therefore urgently necessary to enrich the technical–methodical discussion platforms with such aspects.

3.4 Reducing Complexity by Means of Indicators

3.4.1 Indicators—A Case Study

At the end of this chapter, many of the considerations that have been made will now be applied to an example, a case study. Indicators are particularly suitable for this because they represent a special type and form of statistical information, the significance of which has increased enormously, without this being reflected in a corresponding increase in methodological guidelines or even standards.

The world is full of indicators. No management trainer, no weather moderator in the media, no business journalist gets by without indicators. Managers define goals in the agreements with their employees and measure their success at the degree of goal achievement. Grants are allocated to projects by quantifying their attractiveness by means of key performance indicators. Education, success of medical treatment, hit rates in sport, etc.: the list of indicators could be continued indefinitely. But what is it that has caused this flood of indicators? Apparently, it is a sufficiently vague, yet attractive concept that hides behind this term: "*Etymologically, an indicator, like an index, has to do with pointing. Anatomically, the indicator muscle (extensor indicis) straightens the index finger. Logically, indicators detect, point or measure, but do*

---

[63] Ulrich Beck has emphasised that the struggle among rationality claims follows from a division of the world between experts (in rationality) and non-experts (deviating from rationality) in the non-reflexive modernity (Beck 1998, p. 57).

[64] For a critical analysis see for example O'Neill *Weapons of Math Destruction: How Big Data Increases Inequality and Threatens Democracy* (O'Neil 2016).

[65] This is the title of the following book (Cavanillas et al. 2018).

not explain. An index in the social sciences typically combines or synthesises indicators, as with the 'index of leading economic indicators', which aims to maximize the predictive value of diverse measures whose movements anticipate the rise and decline of general economic activity. A quantitative index or indicator typically cannot measure the very thing of interest, but in its place something whose movements show a consistent relationship to that thing. Since its purpose is merely to indicate as a guide to action, ease of measurement is preferred to meaning or depth" (Porter 2015: Summary).

The fact that indicators are everywhere and used by different communities for their purposes results in a great diversity in the terminology used. In addition, the integration of new data sources (big data) and new disciplines (e.g. data science) has added even more confusion. Differences between 'data', 'statistics', 'information', 'indicators' or 'metrics' are largely blurred. All these terms are treated as synonyms, and all come together under the umbrella term and buzzword 'data'.

The situation in statistics, however, does not necessarily look much better in general. While indicators have found their place as a valuable tool for the dissemination and communication of statistics,[66] not much attention is devoted yet to them in terms of harmonised methodology, applicable equally as a standard to all statistical domains.

A couple of important agreements concerning indicator methodology do already exist, however. Just to mention a few, the statistical domain of social indicators can look back on a tradition of methodological discussion, conceptual harmonisation and practical implementation since the 1980s[67]; composite indicators have been developed and methodologically harmonised[68]; Eurostat has published a series of guidelines, which could serve as a first step into the direction of a standardised methodology.[69]

Nevertheless, it can be concluded that there is discrepancy between the large dynamics in the quantity and variety of indicators and the comparatively low degree of methodological harmonisation, which implies some not insignificant risks:

- Inflation and proliferation of indicators might lead to misperception in such a way that it does not require the professional treatment of statisticians, but that anyone without proper training of techniques and methods can handle indicators.
- Anecdotal, spontaneous design of indicators might be perceived as appropriate, if it is all too easy to ask for the provision of indicators in the preparation of decisions or when political negotiations are difficult. In such situations, a statistician should first be consulted before any further requests are manifested. The costs (production costs, response burden) and benefits of additional indicators should be considered as part of the impact assessments for regulations and political projects.

---

[66]See, for example, the large selection of '*EU Policy Indicators*' provided by Eurostat (http://ec.europa.eu/eurostat/).

[67]See the *Handbook on Social Indicators* (United Nations 1989).

[68]See *Handbook on Constructing Composite Indicators – Methodology and User Guide* (OECD and EuropeanCommission_JRC 2008).

[69]See the *Guidelines for indicators, parts 1–3* (Eurostat 2014, 2017a, b).

- Fragmentation of methods and terminology and unnecessary variation create confusion on both sides, producers and users of indicators.
- Unrealistic expectations from the political and management level result in frustration, misinterpretation and conflicts. Adaptations in practice 'solve' those problems at the expense of quality.
- Political influence might dominate the design, production and communication of indicators, so that statistical principles and the quality of indicators are threatened.

3.4.2 Methodology for Indicators

The essential lesson from all the above criticisms is that indicators have a very specific role to play in the statistical information portfolio and that, consequently, a specific methodology has to be applied. The methodology used here needs, however, to be tailored for this particular type of information and its utilisation. Ensuring that indicators are not politically driven, albeit policy-relevant, is ultimately the crucial issue on which the decision between trust and mistrust depends. In this respect, questions of communication and governance must be anchored from the outset in the methodology.

Of course, quantitative indicators are statistics. Not all statistics are, however, indicators. In Sect. 2.1.5, indicators were defined as follows: an indicator is a summary measure related to a key issue or phenomenon and derived from a series of observed facts. Indicators can be used to reveal relative positions or show positive or negative change. Indicators are usually a direct input into national, EU and global policies. In strategic policy fields, they are important for setting targets and monitoring their achievement.

Indicators are a special type of statistical information in two respects:

- First, it is a particularly high level of compression, focus and synthesis with regard to the relevant statistical message;
- Second, it is the close connection to a scope, a purpose and the associated user community.

In contrast to the broad and detailed basic statistics as well as the consistent accounts; therefore, indicators have a different quality profile (see Fig. 3.2) with a particular emphasis on the criterion of relevance. They have a specific job to do, namely to condense and communicate the informational content contained in statistics in such a way that it can be understood and used by the respective target group. Indicators are pointers that point to particular feature. In contrast to multipurpose basic statistics, indicators are designed (or at least should be designed) in such a way that they serve one (partly very specific) purpose.

It is now crucial to consider the two peculiarities and dimensions of indicators as interrelated and interdependent. A special focus, aggregation and consolidation of various pieces of information into one indicator is only possible if it is known for what purpose it is planned to use that indicator. Even an already highly condensed aggregate of National Accounts, such as GDP, requires further specifications in order to be customised as an indicator to suit a specific application: shall inflation be filtered out, shall an index for growth be calculated, is it a net amount without capital depreciation, or is it a seasonally adjusted quarterly GDP that is expected?

Depending on the phase in the life cycle of a policy area targeted by the indicator, different characteristics are expected and needed. In a phase of awareness raising for a new phenomenon, it might, for example, be sufficient to work with indicators of lower accuracy and granularity while for the monitoring of target achievement, very high precision and resolution is a necessary feature of provided indicators.

Indicators can reveal, suggest, distort and conceal.[70] Which of these characteristics is actually built in the design of an indicator or an indicator set will determine their impact on the debate in the 'bazaar'.[71]

It follows that it is neither meaningful nor possible to develop indicators external to and isolated from the system that is going to use this information, in a separate area of statistics or solely through collaboration of statistics with science. Rather, a co-construction is required, where different stakeholders contribute and participate. From a systemic point of view, the observing and observed systems cannot be isolated from each other.

Taking note of this fact leads consequently to different questions, expectations and approaches concerning also an envisaged indicator methodology. As pointed out in the example of Sustainable Development (see Sect. 3.5), it is not a question of carrying out a 'measurement of sustainability' as a mere academic and analytical undertaking, which is then fed into the political discussion. Rather, the point is— and this is demonstrated by the current process at the global level—to develop and constantly improve in close connection between the political and statistical worlds the measurement and use of these measurements. The particular challenge is, here, to ensure the quality of indicators (see quality principles and Code of Practice in earlier sections) when they are produced under these circumstances. It must therefore be the goal to ensure both this quality and convince the users of this quality in communication.

Figure 3.10 outlines such a co-construction between a 'laboratory' (collaboration of statistics and science) and a 'bazaar' that interact and have mutual relationships. It is the task of the actors in the laboratory, to reduce the complexity as far as is possible with the technical and methodical tools at their disposal. However, the (pre-)selection of relevant aspects, or the setting of indicator-related targets, or the definition of weighting schemes (as part of composite indicators and rankings) may then, at least to a large extent, belong to the field of politics, i.e. the 'bazaar'. Thus, a reduction of complexity requires inputs from both sides.

In this co-construction, the laboratory (statistics and science) has two tasks: firstly, to develop and apply methods of aggregation and synthesis; and secondly,

---

[70] See Lehtonen (2015), Ravetz et al. (2018).

[71] In this sense, indicators are of high relevance for human rights and (global) governance (Merry 2011).

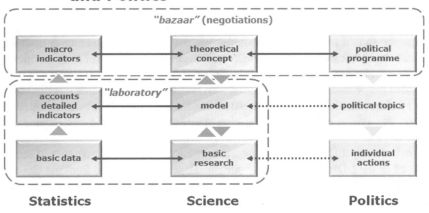

Fig. 3.10 Co-construction of indicators (Radermacher 2005)

to develop and apply methods of active engagement with the users. The decisive factor is that these two tasks are thought, conceived and implemented together, by an interdisciplinary cooperation, as explained in Sect. 3.2.3.

Methodological development and research in the field of indicators is spread over different disciplines, each of them focusing on their respective field of expertise. Furthermore, specific methodological approaches and concepts have been developed side by side in single statistical communities for various sets of indicators. While economics statistics, given that their observation units are often expressed in monetary terms resulting from market transactions, prefer an accounting approach to obtain highly aggregated economic indicators, such as inflation, growth, or productivity, this is difficult to do similarly in social and environmental statistics, where many essential variables are not monetised on actual markets.[72] This is one of the reasons why statistical methods of synthesis and aggregation come into play.[73] Mutual fertilisation through methodical cooperation across the disciplines should therefore be supported—not least because of new opportunities arising from new data sources, the interest in successful methodologies in neighbouring areas (e.g. geographical sciences and statistics) or new disciplines (e.g. data science) as a source of innovation and efficiency.

In parallel with the statistically/methodologically oriented fields, a growing group of researchers has been dealing with the sociological aspects of co-construction of indicators (general or specific ones) in recent years, pointing out the particular challenges of indicators in terms of communication as boundary objects or concerning

---

[72]Even if this might change, following the analysis of Cook (2017).

[73]For an overview over the statistical methods and approaches, see Maggino's *Complexity in Society: From Indicators to their Synthesis* (Maggino 2017).

required governance provision.[74] Which function, for instance, should indicators have in the political process? Should they be used to facilitate an exchange of views ('opening-up') or should they be used to shorten or close a discussion ('closing-down')? What is the relationship between targets, goals and indicators? In which sequence should they be defined and by whom? How much evidence is available from market research for the (correct) effectiveness of indicators? Is this evidence systematically fed into the learning cycles of statistics? To what extent and at what stage of the process is a consultation or even stakeholder participation (i.e. civil society) taking place? Such questions would require significantly intensified research work, and this in collaboration between statisticians, communication experts and sociologists.

Indicators with a political weight, such as the price index, the number of unemployed or the GDP, did not arise overnight. Their strength stems from the fact that they have evolved over years and decades, in a constant interplay between new user demands, conceptual advancements based on scientific work and new data sources, and statistical methods. Therefore, it cannot be expected that, for complex issues (such as Sustainable Development), an indicator or system of indicators can be born with a forced act, even if the political will so demands. It is more reasonable to assume that such a system evolves and learns, that all participants (science, statistics, society) contribute to this evolution, and that this long-term and multifaceted process requires order, governance and management.

Indicators with a high political impact, such as the indices of indebtedness of the public sectors in Europe, which are closely linked to the Excessive Deficit Procedure (EDP, introduced with the common currency by the Maastricht Treaty[75]) belong to an extraordinary category of indicators that deserves a special degree of attention. As history has shown[76] the fact that these indicators are used for surveillance purposes and that they could result in serious consequences contains the risk of manipulation and interference by politics. The special degree of authority that has been assigned to these indicators therefore requires special governance that must go beyond the usual dose of institutional competencies of statistics.

Indicators with a high potential to influence the markets, such as the inflation rate or the quarterly GDP, deserve special attention with regard to equal access and in particular pre-release access.[77] Like many statistical offices, Eurostat has

---

[74]See, for example, the following selection: Power (1994), Desrosières (2010), Sangolt (2010a, b), Desrosières (2011), Hammer (2011), Saetnan et al. (2011), Davis et al. (2012a, b), Coyle (2014), Sébastien et al. (2014), Porter (2015), König (2015), Rottenburg et al. (2015), Supiot (2015a, b), Davies (2016), Diaz-Bone and Didier (2016), Cherrier (2017), Cook (2017), Eyraud (2018), König (2018a, b), Ravetz (2018), Ravetz et al. (2018).

[75]For more details see http://ec.europa.eu/eurostat/web/government-finance-statistics/excessive-deficit-procedure.

[76]See the example of Greek statistics in Sect. 3.2.1.1.

[77]See Eurostat's policies here (http://ec.europa.eu/eurostat/about/policies/dissemination).

discussed this subject with representatives of the media and has recorded the results in a *"Protocol on impartial access to Eurostat data for users"*.[78]

Rankings and indicators with a weighting not based on observations or explicit market values do not fall into the area of official statistics. Nonetheless, it is very useful, through close cooperation between applied research and official statistics, to offer users a range of highly condensed information that combines the best available statistics with the best practices for their compression.

3.4.3 Indicators, Goals, Targets, Monitoring

The process model for co-construction of indicators developed in the previous section is, in principle, not only open to the setting of concrete goals by the politicians, but assumes that this is indispensable for a strong consolidation and synthesis of indicators. Political goals are manifest value judgements of a society. Without being in charge to assess their validity, statisticians can make use of them as a reference for statistical indicators.

As a result, a much simpler and clearer representation of multidimensional indicator sets is possible both numerically and, above all, graphically. For example, Eurostat uses a graphical representation that is based on the theoretical development path between the starting point of the indicator and the target value at the end point in time (see Fig. 3.11).

Nonetheless, there are also critical aspects of a procedure that is geared to target values. Sakiko Fukuda-Parr has addressed the intentional and unintended effects of a target-based policy under the UN Millennium Development Goals (MDGs) (Fukuda-Parr 2015) and similarly questions the Sustainable Development Strategy (SDG) (Fukuda-Parr 2018).

An interplay between the statistical process of the indicator design and the political process of defining objectives can succeed if this is done gradually. First, goals should be distilled out and politically agreed, which subdivide a broad topic and qualitatively formulate the priority questions. In the second step, the statistics are required to provide a set of quantitative answers in the form of indicators for qualitative questions. If it has been agreed that these indicators are suitable, then the third step on the part of the politicians is to fix quantitative targets to be set up at an agreed time in the future. In the fourth step, a monitoring can then begin, in which the indicators and their development are assessed against the targets in order to be able to determine whether the target is being approached as agreed.

If one does not stick to this step sequence but makes, for example, the third before the second step (setting of targets before the indicators are designed and agreed upon), one generates: (a) targets for which there are no indicators; (b) targets that remain

---

[78]See   http://ec.europa.eu/eurostat/documents/4187653/5798057/IMPARTIAL_ACCESS_2014_ JAN-EN.PDF.

(a)

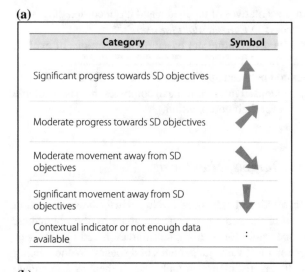

Category	Symbol
Significant progress towards SD objectives	⬆
Moderate progress towards SD objectives	⬈
Moderate movement away from SD objectives	⬊
Significant movement away from SD objectives	⬇
Contextual indicator or not enough data available	⋮

(b)

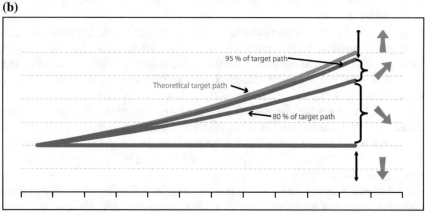

95 % of target path

Theoretical target path

80 % of target path

Fig. 3.11 **a** Trend categories and associated symbols (*Source* Eurostat 2017c, p. 24) and **b** schematic representation of the approach (with quantitative targets) (*Source* Eurostat 2017c, p. 25)

partly qualitative (they are only subdivisions of the priority goal); and (c) the inflating of the number of indicators unnecessarily and with redundancy and inconsistencies.

3.4.4 Lessons Learned for Indicators

These remarks make it clear that it is necessary and useful to work on a comprehensive indicator methodology for official statistics. The momentum generated by the Sustainable Development Indicators (SDI) certainly helps to come considerably closer to this goal. There are important lessons to be learned for the future of indicators in and by official statistics.

Overall, it seems important to reduce the fragmentation of the indicator landscape by consistently working to bring together and bridge developments in different communities. For official statistics, it should be possible to come to a (at least in core) standardised terminology and methodology covering the entire production process from data to the composite indicator. Similarly, in such a comprehensive methodology manual, it is necessary to do justice to the specific nature of the indicators (i.e. boundary objects) by including the components of communication, participation, market research and governance.

Especially in the area of Sustainable Development (SD), it should be possible to use the indicator system for the entire planning and production process (mainstreaming SDI) be it through user-friendly links to basic statistics and accounts or through adaptation of communication and media work by prioritising indicators in long-term planning.

Partnerships with science and neighbouring institutes can help to join forces and increase the effectiveness of the evidence provided.

In terms of governance, it is important to grant statistics the professional degree of freedom it needs to provide quality information. Determining the details of an indicator system, as with the enormously large number of targets in the case of SD, reduces these degrees of freedom sensitively and can ultimately lead to unsatisfactory results.

Specific advice for improving indicators is summarised in the following list:

1. Promote epistemological knowledge and sensitivity among scientists, practitioners and users; promote media and information literacy to counter disinformation

2. Improve evidence about the impact of indicators through market research and impact assessments

3. Promote a focus on quality and be critical with unrealistic expectations

4. Make use of state-of-the-art aggregation methods, minimising normative (value-based weighting) ingredients; be transparent, when normative components (judgements, weights, values) are mixed with observation-based indicators

5. Connect indicators with other components of the information system of official statistics (accounts, basic statistics); make use of the complementarity of statistical tools (in particular accounts)

6. Promote indicator-related research and innovation; promote indicator-related methodological competencies with appropriate training

7. Establish platforms and channels that facilitate communication among scientists, indicator producers and users of indicators; find regular opportunities for indicator-related exchange of views in (international) conferences

8. Ensure a functioning dialogue between goal setting and indicator design processes; assign power to indicators for the redefinition of targets and goals

9. Promote an evolutionary perspective, where continuous improvement of the indicator design is part of the strategy

10. Develop tools for empowering users and journalists to tackle disinformation and foster a positive engagement with fast-evolving information technologies
11. Safeguard the strengths and independence of institutions and researchers working in the area of indicator development, production and knowledge management
12. Enhance transparency about the quality of indicators and compliance with ethical principles and good governance guidelines.

3.5 Sustainable Development

3.5.1 A Simple, Perhaps Too Simple Principle

Sustainable Development means to maximise the welfare of the current generation (now and here) without jeopardising the starting position for future generations (tomorrow) or living at the expense of people in other countries (elsewhere).

This basic principle was already introduced by the Brundtland commission at the end of the 1980s (World Commission on Environment and Development 1987) and adopted by the 'World Summit' in Rio 1992. The implementation was planned to follow the 'Agenda 21', which included a chapter on *"Information for decision-making 40.1: In Sustainable Development, everyone is a user and provider of information considered in the broad sense. That includes data, information, appropriately packaged experience and knowledge. The need for information arises at all levels, from that of senior decision makers at the national and international levels to the grass-roots and individual levels. The following two programme areas need to be implemented to ensure that decisions are based increasingly on sound information: 1. Bridging the data gap; 2. Improving information availability"* (United Nations 1992).

Twenty-two years after the 51st Session of the International Statistical Institute in Istanbul in 1997, many of the statistical questions and answers on the topic of Sustainable Development already presented by the author at the time (Radermacher 1999: 242) are still valid: *"Sustainable Development can only be defined and achieved by a complicated restructuring process of the society including the fact that the final results of that process cannot be anticipated by (scientific) assumptions and (statistical) surveys or estimates. Consequently, the quantitative results of "Environmental Economic Accounting" or "Sustainability Indicators" must be interpreted as representations of possible margins for the manoeuvre to Sustainable Development, and inputs to policy debate in this sense".*

"What is sustainability? How can we make Sustainable Development a reality? How sustainability can be measured?" is a set of questions, which, in the course of

two years, did receive 40,649 reads, 722 followers and 2909 answers in the social net-work ResearchGate.[79] Obviously, these questions are related and cannot be answered separately. It is not possible to measure Sustainable Development independently of the political–social change process that needs to be done to reach this goal. The statistical question behind it is with which metrics the decisions on the way to Sustainable Development can be improved and of what quality and condition these metrics have to be, so that they are robust enough for the social conflicts that have to be solved. The complexity inherent in Sustainable Development must be reduced and simplified, but without going too far or risking the credibility of metrics with covert value judgements. Indicators are a very suitable tool for this. However, for indicators to actually play their intended enlightening role, more is needed than statistical methodology and reliable production of information. In addition, factors must be taken into account which, both in the design and in the communication of indicators, place their functionality and the target group of users in the political–societal context in the foreground.

After an intermezzo with 'Millennium Development Goals', the UN General Assembly in 2015 reiterated the concept of Sustainable Development with its adoption of Sustainable Development Goals and the set-up of the 2030 Agenda (United Nations 2017b), which outlines the way in which the implementation should be monitored: *"At the global level, the 17 Sustainable Development Goals (SDGs) and 169 targets of the new agenda will be monitored and reviewed using a set of global indicators. The global indicator framework, to be developed by the Inter Agency and Expert Group on SDG Indicators (IAEA-SDGs), will be agreed on by the UN Statistical Commission by March 2016. The Economic and Social Council and the General Assembly will then adopt these indicators … Chief statisticians from Member States are working on the identification of the targets with the aim to have 2 indicators for each target. There will be approximately 300 indicators for all the targets. Where the targets cover cross-cutting issues, however, the number of indicators may be reduced"*.

Upon this political commitment, the community of official statisticians is asked to provide appropriate metrics of Sustainable Development, which should quantify not only whether a society is making progress, but also whether this progress will have consequences tomorrow or elsewhere. This is the broadest thinkable concept, containing all the kinds of measurement and quantification challenges that a statistician can imagine. The idea of Sustainable Development, born more than 30 years ago, seems so simple: improve the living conditions of today's generation without worsening the chances of future generations. For 30 to almost 50 years, this objective has seen attempts for it to be qualified (i.e. mirrored in an adequate theoretical model) and quantified (i.e. to provide Sustainable Development metrics as evidence for policy decisions).

---

[79]Extracted on 21.10.2019 https://www.researchgate.net/post/What_is_sustainability_How_can_we_make_sustainable_development_a_reality_How_sustainability_can_be_measured.

Since five decades, the development of environmental metrics followed several historical episodes (and corresponding measurement tools), which were driven by environmental crises:

- first, volume and price changes: energy crisis, (quantitative) depletion of natural resources;
- second, local/regional changes of quality: air, waste, water, etc., (qualitative) degradation;
- third, global phenomena such as ozone layer, climate change;
- fourth, ecosystems, biodiversity and planetary boundaries.

We are now in this fourth episode, which is moving beyond the measure of individual stocks of natural capital and towards ecosystems, which constitutes the interplay of different systems. This definition makes clear that in order to measure sustainability in the context of the environment, a movement beyond the measurement of stock is required. These ecosystems:

- are not only (quantitative) stocks but dynamic systems and, as such, they can have greater or lesser degrees of resilience;
- provide a multitude of services to a society (e.g. a forest not only supplies timber but may also provide water retention and flood or landslide protection, air filtration, carbon sequestration, habitat for rare species and recreation).

The interrelationships between nature, the economy and social processes must themselves play a role in the design of the corresponding metrics of the ecological, social and economic systems. This is basically the case when applying the heuristics and objectives of Sustainable Development. At the same time, the question of how, after five decades of environmental policy and environmental statistics, we succeed in actually approaching the goal of Sustainable Development at the pace required in view of the dramatic global environmental risks, is becoming increasingly important. In addition, the question arises of how civil societies can be so reliably informed with statistics that they can respond to disruptive manoeuvres on the part of politicians (Sitglitz 2019; Davies 2017, 2018).

3.5.2   Conceptual Approaches from Different Angles

Over this long period, very different approaches have been developed (partly in parallel streams and communities) to bring the interactions between humans and the environment into a logical framework in order to create order and structure in the ocean of possible and actual information. In the meantime, many of these approaches have been incorporated into the System of Environmental-Economic Accounting SEEA (United Nations 2014b) (a UN statistical standard) and its experimental part, which addresses open methodological issues of ecosystem accounting

(United Nations 2014c). In contrast to the System of National Accounts, the SEEA is composed of several methodological approaches under one roof[80]:

- Flow accounts, which ultimately can be traced back to the approaches of interpreting the economy as an organism whose metabolism can be represented in a meaningful way by input–output methods (Ayres and Simonis 1994; Radermacher and Stahmer 1998).
- 'Footprints', which pursue the idea of a cumulation of all environmental pressures (be it the emission of greenhouse gases or the use of land) that arise 'from cradle to grave' for a good, i.e. over the entire production route, including all energy inputs and transport routes, wherever these take place, at home as well as abroad.[81] Input–output analysis methods offer the appropriate tools for this, provided the corresponding input–output tables (e.g. World IoTs) are available. In a globalised world, such statistics are particularly important for quantifying the cross-border effects of production and consumption, i.e. the export and import of environmental pollution.
- Inclusion of the nature (and its services) in the capital accounts to determine whether or not a nation's capital has been kept intact over a period (Radermacher and Steurer 2015; De Smedt et al. 2018). In this way, the economic method of accounting for long-term effects resulting from the use and depreciation of fixed assets is also applied to nature. The future with the interests of the then living generations thus becomes part of today's balance sheet.
- A systematic and consistent presentation of economic activities, such as goods, services and taxes that are directly related to environmental issues. In contrast to the previous modules, the system boundaries of the economy are not extended here. Rather, it is a matter of subdividing existing aggregates with finer granulation in order to be able to analyse environmentally relevant changes over time.
- A monetary valuation of natural goods and their services through methods applied in environmental economics is handled cautiously in SEEA for many reasons. Although this would allow a higher degree of compression and aggregation of information to a few indicators, the inevitable value judgement load is considered to be incompatible with the quality profile of official statistics (Radermacher and Steurer 2015; Cook 2017).

Parallel to the accounting methods described above, basic statistics in the field of the environment have also developed considerably. The corresponding international guideline is the Framework for the Development of Environment Statistics (FDES) of the United Nations (United Nations 2017a). The distinction between Driving forces, environmental Pressures, the State of the environment and Responses of society (DPSR) has played a crucial role in the systematics of environmental statistics as well as in indicators in the environmental field. For a further statistical decomposition

---

[80]See, for example, https://ec.europa.eu/eurostat/statistics-explained/index.php/Environmental_ accounts_-_establishing_the_links_between_the_environment_and_the_economy#Introduction_ to_environmental_accounting.

[81]See, for example, the calculation of Total Material Requirement https://www.eea.europa.eu/ publications/signals-2000/page017.htmlt.

of the different driving forces (population, affluence and technology), the I(mpact on the environment) = P(opulation) × A(ffluence) × T(echnology) 'formula' (Ehrlich and Holdren John 1971) can be used additionally in order to integrate statistical information other than environmental into the picture. Like all alternative approaches mentioned, this 'formula' must not be overinterpreted as a mathematical or physical equation. Rather, it is suited to draw attention to the relationships, modes of interaction and feedback between demographic development, prosperity and technological progress. In this way, it is possible, for example, to place statistical information on demographic developments (global and national) in the context of improvements in living conditions (income, health, education, etc.) (Rosling et al. 2018; Population Matters 2018) and to derive the resulting additional environmental burdens from the use of the existing technologies (production methods, mobility, food, etc.).

Geographic information systems (GIS), geocoded data and geostatistical methods have developed continuously since the late 1980s and offer correspondingly long time series, for example on land cover and land use and their changes over time.[82] What is decisive here is the opening-up of completely new possibilities for linking basic data via their spatial relationship. When considering the accounting of ecosystems and biodiversity because their services are vital for the survival of mankind, it is necessary firstly to record these systems in a statistically representative way (analogous to, for example, population statistics) and secondly to use appropriate statistical methods to aggregate such large amounts of data into meaningful indicators.[83] With the current availability of large sets of data from remote sensing, the first condition will be easier to fulfil than in the past. Nevertheless, remote sensing data must be combined with (costly and cumbersome) results from field surveys in smart GISs if progress is to be made in this area (Radermacher et al. 1998).

The disadvantage of many of the methodological approaches described so far is that they do not adequately reflect the complexity of the interdependencies between economic, social and ecological systems, i.e. they are too reductionist. Therefore, a complementary approach is increasingly being pursued, which deals with these systems, their vulnerability/robustness, any existing tipping points and the risks of crossing such sensitive boundaries. *"In a systems approach, the focus moves from measuring the stocks of assets to coming to grips with the resilience of economic, societal and natural systems. Tackling these issues requires interdisciplinary work, with a focus on the ability of the system to cope with risks and uncertainties in a broad and long-run perspective, and on the different ways to manage this coping*

---

[82]See, for example, the archive of European data from CORINE Land Cover https://land.copernicus.eu/pan-european/corine-land-cover or of LUCAS https://esdac.jrc.ec.europa.eu/projects/lucas and https://ec.europa.eu/eurostat/statistics-explained/index.php/LUCAS_-_Land_use_and_land_cover_survey.

[83]In Europe, corresponding activities are coordinated within 'the Mapping and Assessment of Ecosystems and their Services (MAES)' https://biodiversity.europa.eu/maes, supported by the knowledge innovation project on an 'Integrated System of Natural Capital and Ecosystem Services Accounting in the EU (INCA)' http://publications.jrc.ec.europa.eu/repository/handle/JRC110321. At UN level, experimental ecosystem accounting has made progress as well (https://seea.un.org/events/forum-experts-seea-experimental-ecosystem-accounting).

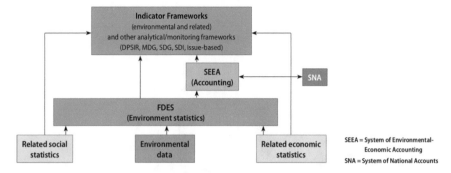

Fig. 3.12 Relationship of basic statistics, accounting systems and indicator sets. From Framework for the Development of Environment Statistics (FDES 2013), by UN Department of Economic and Social Affairs—Statistics Division, ©2017 United Nations. Reprinted with the permission of the United Nations (2017a: 25)

ability (resilience) of systems. "Resilience" is indeed referred to in the Sustainable Development Goals (SDGs) and by the targets of the 2030 Agenda" (De Smedt et al. 2018).

Indicator approaches differ fundamentally from the methods mentioned before (see Fig. 3.12). The main purpose of indicators is usually to define quantitative measures for a particular policy direction or programme in such a way that these 'metrics' can also be used to set quantitative targets and to monitor the achievement of objectives. While basic statistics and accounts are multipurpose, indicators are closely linked with a specific application. This makes a huge difference in the way indicators have to be designed, produced and communicated. Indicators are not a simple subcategory of statistics. With their earmarking, completely different driving forces and intended functionalities come into play, which must be taken into account.

Above all, however, it is important to understand indicators as one product in the portfolio of official statistics. Indicators must be closely linked to, build on and lead to the other products for further analysis (see Fig. 3.13).

The 2014 Recommendations of the Conference of European Statistics give a comparatively complete overview of the progress made in the areas of human well-being 'here and now', 'later' and 'elsewhere'. Noteworthy is the assessment in the supposedly simplest of these three dimensions, the 'here and now': *"There is no theoretical consensus on how to measure the human well-being of the present generation"* (UNECE 2014, p. xviii).

In the theoretical and statistical analysis of interactions between the natural, the social and the economic systems in the sense of nonlinear and erratic changes, resilience or vulnerability, planetary boundaries, etc., the development of corresponding metrics is comparatively at the beginning[84] and a comprehensive standard

---

[84]The Report of the High-Level Expert Group on the Measurement of Economic Performance and Social Progress (HLEG) was released during the 6th OECD World Forum on Statistics, Knowledge and Policy on 27–29 November 2018 in Incheon, Korea; this aspect is covered in the report https://www.oecd.org/statistics/measuring-economic-social-progress/ (Stiglitz et al. 2018b).

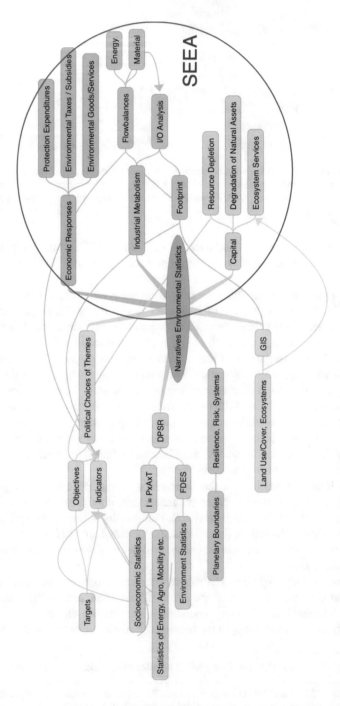

Fig. 3.13 Narratives, heuristics and frameworks used in statistics of the environment

of statistical measurement (if at all possible) for these domains is still a long way off, even if presumably comprehensive indicator lists are floating around (O'Neill et al. 2018).

3.5.3 From Theoretical Concepts to the Production of Qualitative Statistics

In reality, Sustainable Development metrics suffer from essential problems of statistical capacity (whether in terms of financial or human resources) or governance (independence of statistics, political standing) or unavailability of basic data. Various initiatives have been launched over the past few years, not least to address the particular handicaps of developing countries.[85]

What currently has the greatest impact (with both opportunities and threats) on development is the United Nations decision to focus long-term strategy for 2030 on Sustainable Development. On the one hand, the status of official statistics as a supplier of the necessary evidence was upgraded. On the other hand, the result of the global political negotiation process with a large number of 'Goals' (17) and an immense number of 'Targets' (169) leaves too little scope for the adequacy process of statistics.[86] Moreover, the international statistical cooperation in the search for standards of SD metrics is significantly influenced by the politically agreed primacy given to individual nation states as process-owners. It remains to be seen to what extent it will be possible to statistically approach the younger but nevertheless strategically essential global issues, such as 'planetary boundaries', under these settings.[87]

The use and influence of indicators as an approach to measure or (better) 'quantify' Sustainable Development has been analysed by a few scholars, who are interested in the potential effectiveness of indicators as 'boundary objects' in connecting science and policy (Sébastien et al. 2014) or who observe this new global approach comparing with its predecessors (Fukuda-Parr et al. 2014; Fukuda-Parr 2017). In comparison with a science-driven approach (i.e. accounting), an indicator approach comes along with very special advantages and shortcomings. To be aware of these is a necessary condition for a mitigation of corresponding risks for the statistical community as well as for the wider public.

---

[85] Data initiatives opening official statistics for data sciences and other (big) data sources (SDG16 2017; PARIS21 2017; United Nations 2014a, b; UNSD 2017).

[86] See https://sustainabledevelopment.un.org/post2015/transformingourworld.

[87] An overview of the policy-driven activities and their critique can be found in Fukuda-Parr (2017), United Nations (2016), Development (2017), SDSN (2017), United Nations 2017a).

3.5.4   Lessons Learnt

But what is the conclusion and what lesson can be learned from this example? Complexity in the case is compounded by the complexity brought about by the necessary cooperation of completely different disciplines/communities, scientific as well as political and statistical. Areas that concern the 'here' and 'now' should be merged with those that encompass the 'tomorrow' or 'elsewhere'. Expectations that there will be a single, clear-cut statistical standard solution for the question of SD metrics neglect the complexity and dynamic nature of the undertaking. Are we acting sustainably? How big is the sustainability gap of my country? Does the society of my country make any progress on balance? For all these questions, there is no such thing as the answer to the question about the height of my child.[88]

> Paul Cilliers has summarised his view in the following way: *"This is no surprise if one grants the argument that a model of a complex system will have to be as complex as the system itself. Reduction of complexity always leads to distortion. ...it problematizes any notion that data can be transformed into knowledge through a pure, mechanical, and objective process. However, it also problematizes any notion that would see the two as totally different things. There are facts that exist independently of the observer of those facts, but the facts do not have their meaning written on their faces. Meaning only comes to be in the process of interaction. Knowledge is interpreted data."* (Cilliers 2000, p. 10)

If it is not an option to provide (ex ante and external to politics) a solid and comprehensive balance sheet that could tell us how far away we are from the desired state of SD or how much we have moved in that direction, then what does statistical evidence has to offer?

The path that societies and states will take on their way to the strategic goals in 2030 will not be straightforward, linear or undisrupted. Rather, it will be a '*transformative social learning process*' (König 2018b), a manoeuvre with many stakeholders and interests, acquaintances and unknowns, a political 'bazaar'. Negotiations and decisions in this 'bazaar' will have to be made at all levels and in many forms. All of these decisions should be based on the best possible 'metrics' from the laboratories of science and factories of statistics available at that time. Broad participation of the civil society should benefit from the open access to all relevant information.[89] Indicators alone do not provide any assistance in the selection and implementation of the necessary measures. Thus, it is probable that continuing current policies without a proper change in lifestyles will result in achieving social goals while violating the boundaries of global ecosystems (Randers et al. 2018).

---

[88]42 is the 'Answer to the Ultimate Question of Life, the Universe, and Everything' in *The Hitchhiker's Guide to the Galaxy* (Adams 1981).

[89]"*Measuring progress may be one important way to renew democracies in decline. In communities around the world, engaging citizens in helping to define and measure progress – a meaningful task which necessarily involves developing a shared vision, identifying concrete outcomes and discussing differences – has proved an important means of rebuilding democratic capacity at a time when many countries show evidence of a general decline in democratic confidence and vitality, as well as alienation and disaffection among their citizens*" (Hall and Rickard 2013, p. 26).

Whether statistical information can do this job depends not least on the fact that all stakeholders are aware of both the possibilities and the limits. If official statistics are so closely linked to the political sphere (like in the indicator approach chosen), it is necessary to face this fact and to ensure with good governance that statistics are policy-relevant but not politically driven. *"Despite, or perhaps because of the addiction to measurement, the pitfalls of counting are frequently under-communicated, by decision-makers, researchers, and perhaps not least by the media. Numbers are too often taken at face value"* (Sangolt 2010a, p. 102). Criteria for the quality of SD metrics can help to avoid unrealistic expectations and facilitate a fruitful and undisturbed user-producer communication (see Box 3.2).

We have shown in this chapter that statistical *"knowledge generation results from making 'simplifying' assumptions. If these are all true, we have a truly appropriate and useful model; but if the assumptions do not hold, our model may be completely misleading"* (Allen 2000, p. 100). Parallel to each other and intertwined in a 'co-production' social conditions and official statistics evolve in a learning process. From this evolution, new results continuously emerge, which have learned from the past, from experiences and from strengths or weaknesses of previous solutions, but which also have to be addressed to the new questions of the time.

This finding is far from being a new discovery for official statistics. Rather, this evolution of programme, methods, products and services is a characteristic feature of official statistics, recognisable across the very different eras of the past 200 years of history.

Box 3.3 Sustainable Development

Better metrics—Better decisions: Main principles

Sustainable Development means (in particular from an environmental perspective) living well within the boundaries of the planet.

Better metrics should enable better decision-making in favour of SD.

To support better decisions in favour of sustainability, metrics should:

- capture the complexity of the phenomena at stake (resilience, risks…);
- be state of the art (multiple sources, technologies, models…);
- be of adequate quality (coverage, timeliness, comparability across space and time) and be relevant (fit for purpose in policy life cycle) by applying a comprehensive quality framework.

'Better' are decisions if they enable societies to navigate through unknown territories towards a more sustainable way of living. This means that today's challenges and risks (e.g. through globalisation) need to be addressed in such a way that democratic opinion making is supported through participation and communication.

SD metrics are continuously being improved in a co-evolution (i.e. a collaborative, iterative learning process) with societal changes on the way to the long-term goal of Sustainable Development. This evolutionary perspective should be enhanced by multiple iteration in learning cycles and interaction between different actors and disciplines:

a. scientific theory/models—empirical data/models;
b. data/information—use/application;
c. scientific theory/models—use/application.

The learning and continuous improvement should be based on 'evidence of evidence' (empirical analysis of the impacts of SD metrics).

Partnership, trust and governance are essential success factors; one should:

- establish platforms and channels that facilitate communication among scientists, producers and users of metrics;
- develop tools for empowering civil society to tackle disinformation, foster engagement with fast-evolving information technologies;
- safeguard the strengths/independence of institutions/researchers working in the area of data sciences/statistics;
- enhance transparency about the quality of information and compliance with ethical principles and good governance.

References

Ackoff, Russell. 1994. From Mechanistic to Social Systemic Thinking. In *Systems Thinking in Action Conference*.

Adams, Douglas. 1981. *The Hitchhiker's Guide to the Galaxy*. Pocket Books.

Allen, Peter M. 2000. Knowledge, Ignorance, and Learning. *Emergence, Complexity and Organisation* 2: 78–103.

Andrews, Tom. 2012. What is Social Constructionism? In *Grounded Theory Review*. Mill Valley: Sociology Press.

Ayres, Robert U., and Udo Ernst Simonis. 1994. *Industrial Metabolism: Restructuring for Sustainable Development*. Tokyo, New York: United Nations University Press.

Barbieri, Giovanni A. 2018, forthcoming. Statistics, Reality, Truth. In *10th ICOTS Conference*. Kyoto.

Beck, Ulrich. 1998. *Risk Society Towards a New Modernity*. London: Sage.

Benessia, A., S. Funtowicz, M. Giampietro, A. Guimaraes Pereira, J. Ravetz, R. Strand, A. Saltelli, and J.P. van der Sluijs. 2016. *The Rightful Place of Science: Science on the Verge*. Tempe, AZ: Consortium for Science, Policy and Outcomes.

Bowker, Geoffrey C., and Susan Leigh Star. 2000. *Sorting Things Out Classification and Its Consequences*. Cambridge, MA: The MIT Press.

Box, George E.P. 1976. Science and Statistics. *Journal of the American Statistical Association* 71: 791–799.

Brandolini, Andrea. 2016. The Links Between Household Surveys and Macro Aggregates. In *DGINS Conference 2016*, ed. Statistics Austria. Vienna: Statistics Austria.

Bröckling, Ulrich, Susanne Krasmann, Thomas Lemke, and Michel Foucault. 2000. *Gouvernementalität der Gegenwart: Studien zur Ökonomisierung des Sozialen*. Frankfurt am Main: Suhrkamp.

Brown, W. 2015. *Undoing the Demos: Neoliberalism's Stealth Revolution*. Cambridge, MA: MIT Press.

Bubrowski, Helene. 2017. Das geschätzte Volk. *Frankfurter Allgemeine* 24 (10): 2017.

Bundesverfassungsgericht. 1983. BVerfG · Urteil vom 15. Dezember 1983 · Az. 1 BvR 209/83, 1 BvR 484/83, 1 BvR 420/83, 1 BvR 362/83, 1 BvR 269/83, 1 BvR 440/83 (Volkszählungsurteil), ed. Bundesverfassungsgericht. Karlsruhe: openjur.

Bundesverfassungsgericht. 2017. Mündliche Verhandlung in Sachen "Zensus 2011" am Dienstag, 24. Oktober 2017, ed. Bundesverfassungsgericht. Karlsruhe: Bundesverfassungsgericht.

Bundesverfassungsgericht. 2018. Zensus 2011 - 2 BvF 1/15 - Rn. (1-357). In *Bundesverfassungsgericht*. Bundesverfassungsgericht.

Burchell, Graham, Colin Gordon, Peter Miller, and Michel Foucault. 1991. *The Foucault Effect: Studies in Governmentality: With Two Lectures by and an Interview with Michael Foucault*. London: Harvester Wheatsheaf.

Cassata, Francesco. 2017. Eugenics Archive—Italy. Social Sciences and Research Council of Canada. http://eugenicsarchive.ca/discover/world/530b9b5a76f0db569b000010.

Cavanillas, José María, Edward Curry, and Wolfgang Wahlster (eds.). 2018. *New Horizons for a Data-Driven Economy—A Roadmap for Usage and Exploitation of Big Data in Europe*. Springer Nature.

Cherrier, Beatrice. 2017. The Making of Economic Facts: A Reading List. In *The Undercover Historian—Beatrice Cherrier's blog*.

Cilliers, Paul. 2000. Knowlege, Complexity, and Understanding. *Emergence, Complexity and Organisation* 2: 7–13.

Clouet, Hadrien. 2015. Le "chômage BIT": comparaison facile, comparaison fragile? In *Sozialstaat/État Social*, ed. Saisir L'Europe. Berlin/Paris: Saisir l'Europe.

Cook, Eli. 2017. *The Pricing of Progress—Economic Indicators and the Capitalization of American Life*. Harvard: Harvard University Press.

Coyle, D. 2014. *GDP: A Brief but Affectionate History*. Princeton: Princeton University Press.

Coyle, D. 2015. *GDP: A Brief but Affectionate History*. Princeton University Press.

Daly, Herman E. 1987. A.N. Whitehead's Fallacy of Misplaced Concreteness: Examples from Economics. *Journal of Interdisciplinary Economics* 2: 83–89.

Dasgupta, Rana. 2018. The Demise of the Nation State, *The Guardian*, Thu April 5, 2018.

Davies, William. 2016. *The Limits of Neoliberalism—Authority, Sovereignty and the Logic of Competition*. London: SAGE Publications.

Davies, William. 2018. *Nervous States—How Feeling Took Over the World*. London: Vintage Publishing.

Davies, William. 2017. How Statistics Lost Their Power—And Why We Should Fear What Comes Next. *The Guardian*.

Davis, Kevin E., Angelina Fisher, Bendict Kingsbury, and Sally Engle Merry. 2012a. *Governance by Indicators—Global Power Through Quantification and Rankings*. Oxford: Oxford University Press.

Davis, Kevin E., Benedict Kingsbury, and Sally Engle Merry. 2012b. Introduction: Global Governance by Indicators. In *Governance by Indicators: Global Power Through Quantification and Rankings*, ed. Kevin E. Davis, Benedict Kingsbury, and Sally Engle Merry. Oxford: Oxford University Press.

De Smedt, Marleen, Enrico Giovannini, and Walter J. Radermacher. 2018. Measuring Sustainability. In *For Good Measure: Advancing Research on Well-being Metrics Beyond GDP*, ed. Joseph E. Stiglitz, Jean-Paul Fitoussi, and Martine Durand. Paris: OECD Publishing.

De Michelis, Alberto, and Alain Chantraine. 2003. *Memoirs of Eurostat—Fifty Years Serving Europe*. Luxembourg: Publication Office of the European Union.

Deming, W.E. 2000. *The New Economics: For Industry, Government, Education*. Massachusetts Institute of Technology, Center for Advanced Engineering Study.

Desrosières, Alain. 1998. *The Politics of Large Numbers—A History of Statistical Reasoning*. Cambridge, MA: Harvard University Press.

Desrosières, Alain. 2001. How Real Are Statistics? Four Possible Attitudes. *Social Research* 68: 339–355.

Desrosières, Alain. 2002. Adolphe Quetelet. *Courrier des statistiques* 104: 3–8.

Desrosières, Alain. 2010. A Politics of Knowledge-Tools—The Case of Statistics. In *Between Enlightenment and Disaster*, ed. Linda Sangolt. Brussels: P.I.E. Peter Lang.

Desrosières, Alain. 2011. Words and Numbers—For a Sociology of the Statistical Argument. In *The Mutual Construction of Statistics and the Society*, ed. Ann Rudinow Saetnan, Heidi Mork Lomell, and Svein Hammer. New York: Routledge.

DGINS. 2016. Vienna Memorandum. In *DGINS Conference 2016*, ed. Statistics Austria. Vienna: Statistics Austria.

Diaz-Bone, Rainer, and Emmanuel Didier (eds.). 2016. *Conventions and Quantification—Transdisciplinary Perspectives on Statistics and Classifications.*

Dixson-Declève, Sandrine, Jørgen Randers, and Anders Wijkman. 2018. The Club of Rome to William Nordhaus and the Nobel Committee: "Pursue Profitability—Even at the Cost of the Planet?!". Zurich: Club of Rome.

Dreyblatt, A., and E. Blume. 2006. *Innocent Questions*. Consortium Book Sales & Dist.

Ehrlich, Paul R., and John P. Holdren. 1971. Impact of Population Growth. *Science* 171: 1212–1217.

Eugenicsarchive. 2018. Eugenics Archives—What Sorts of People Should There Be? Social Sciences and Humanities Research Council of Canada, Accessed 28.05.2018. http://eugenicsarchive.ca/.

European Commission. 2010. *Report on Greek Government Deficit and Debt Statistics*. Brussels: European Commission.

European Commission. 2018. Report from the Commission to the European Parliament and the Council on the Quality of Fiscal Data Reported by Member States in 2017. Brussels: European Commission.

Eurostat. 2014. *Part 1—Indicator Typologies and Terminologies*. Luxembourg: Eurostat.

Eurostat. 2015a. Annual National Accounts—How ESA 2010 Has Changed the Main GDP Aggregates. Eurostat, Accessed 23.04.2018. http://ec.europa.eu/eurostat/statistics-explained/index.php/Annual_national_accounts_-_how_ESA_2010_has_changed_the_main_GDP_aggregates.

Eurostat. 2015b. Quality of Life in Europe—Facts and Views—Overall Life Satisfaction. Eurostat, Accessed 23.04.2018. http://ec.europa.eu/eurostat/statistics-explained/index.php/Quality_of_life_in_Europe_-_facts_and_views_-_overall_life_satisfaction.

Eurostat. 2017a. *Part 2—Communicating Through Indicators*. Luxembourg: Eurostat.

Eurostat. 2017b. *Part 3—Relevance of Indicators for Policy Making*. Luxembourg: Eurostat.

Eurostat. 2017c. Sustainable Development in the European Union—2017 Edition. In *Statistical Books*, ed. Eurostat. Luxembourg: Eurostat.

Eurostat. 2017d. Unemployment Statistics and Beyond. Eurostat, Accessed 10.04.2018. http://ec.europa.eu/eurostat/statistics-explained/index.php?title=Unemployment_statistics_and_beyond.

Ewald, François. 1991. *Spiele der Wahrheit Michel Foucaults Denken*. Frankfurt am Main: Suhrkamp.

Eyraud, Corine. 2018. Stakeholder Involvement in the Statistical Value Chain: Bridging the Gap Between Citizens and Official Statistics. In *Power from Statistics: Data, Information and Knowledge—Outlook Report—2018 Edition*, ed. Eurostat. Luxembourg: Publication Office of the European Union.

Foucault, Michel. 1991. Governmentality. In *The Foucault Effect*, ed. Graham Burchell, Colin Gordon, and Peter Miller. Chicago: Chicago University Press.

Foucault, Michel. 1978. «La governamentalità» («La gouvernementalité»). Accessed 24.08.2018. http://1libertaire.free.fr/MFoucault136.html.

Fremdling, Rainer. 2016. Zur Bedeutung nationalsozialistischer Statistiken und Statistiker nach dem Krieg - Rolf Wagenführ und der United States Strategic Bombing Survey (USSBS). *Jahrbuch für Wirtschaftsgeschichte* 57: 589–613.

Fried, Samantha J. 2014. *Quantify This: Statistics, the State, and Governmentality*. Georgetown University.

Fukuda-Parr, Sakiko. 2015. Global Goals as a Policy Tool: Intended and Unitended Consequences. In *The MDGs, Capabilities and Human Rights*, ed. Sakiko Fukuda-Parr and Alicia Ely Yamin. New York: Routledge.

Fukuda-Parr, Sakiko. 2017. *United Nations High Level Political Forum Opening Panel, July 10, 2017—Statement by Sakiko Fukuda-Parr*. New York: United Nations.

Fukuda-Parr, Sakiko. 2018. Is goal setting a good way to define global development agendas? In *European Commission Workshop the Impacts and Methodology of Indicators and Scoreboards*. Ispra, Italy: Joint Research Center.

Fukuda-Parr, Sakiko, Alicia Ely Yamin, and Joshua Greenstein. 2014. The Power of Numbers: A Critical Review of Millennium Development Goal Targets for Human Development and Human Rights. *Journal of Human Development and Capabilities: A Multi-Disciplinary Journal for People-Centered Development* 15: 1–13.

Funtowicz, Silvio O., and Jerome R. Ravetz. 1993. Science for the Post-normal Age. *Futures* 25: 739–755.

Gelfert, Axel. 2016. *How to Do Science with Models—A Philosophical Primer*. Switzerland: Springer.

GfdS. 2016. «GfdS wählt» postfaktisch «zum Wort des Jahres 2016», ed. Gesellschaft für Deutsche Sprache. Wiesbaden.

Goldsmiths. 2018. Arithmus—Peopling Europe: How Data Make a People. Goldsmiths—University of London. Accessed 28.05.2018. http://arithmus.eu/.

GreatBritain. 1981. Statistical Services in the Civil Service Department: Report by the Rayner Survey Officer and Statement of Decisions by Ministers: Rayner Review of Government Statistical Services, ed. Civil Service Department. London ([Whitehall, SW1A 2AZ]): The Department.

Grohmann, Heinz. 1985. Vom theoretischen Konstrukt zum statistischen Begriff - Das Adäquationsproblem. *Allgemeines Statistisches Archiv* 69: 1–15.

Gueye, Gallo. 2016. Closing gaps and producing official statistics on Income, Consumption and Wealth (ICW). In *DGINS Conference 2016*, ed. Statistics Austria. Vienna: Statistics Austria.

Hacking, Ian. 1991. How Should We Do the History of Statistics? In *The Foucault Effect—Studies in Governmentality*, ed. Graham Burchell, Colin Gordon, and Peter Miller. Chicago: University of Chicago Press.

Hall, Jon, and Louise Rickard. 2013. *People, Progress and Participation—How Initiatives Measuring Social Progress Yield Benefits Beyond Better Metrics*. Gütersloh: Bertelsmann Stiftung.

Hammer, Svein. 2011. Governing by Indicators and Outcomes: A Neo-liberal Governmentality? In *The Mutual Construction of Statistics and Society*, ed. Ann Rudinow Saetnan, Heidi Mork Lomell, and Svein Hammer. New York: Routledge.

Hand, David J. 2009. Modern Statistics: The Myth and the Magic. *Journal of the Royal Statistical Society* 2009: 287–306.

Hendricks, Vincent F., and Mads Vestergaard. 2018. *Postfaktisch - Die neue Wirklichkeit in Zeiten von Bullshit, Fake News und Verschwöruungstherien*. München: Karl Blessing Verlag.

Horn, David G. 1994. *Social Bodies—Science, Reproduction, and Italian Modernity*. Princeton University Press: Princeton.

Hufe, Paul, Ravi Kanbur, and Andreas Peichl. 2018. Measuring Unfair Inequality: Reconciling Equality of Opportunity and Freedom from Poverty. In *CESifo Working Paper No. 7119*, 1–47. Munich: Munich Society for the Promotion of Economic Research-CESifo GmbH.

Hunter, John. 2015. Myth: If You Can't Measure It, You Can't Manage It. In *The W. Edwards Deming Institute Blog*, ed. The Deming Institute. The Deming Institute.

Jasanoff, Sheila. 2004a. *States of Knowledge: The Co-production of Science and the Social Order*. London: Routledge.

Jasanoff, Sheila (ed.). 2004b. *States of Knowledge: The Co-production of Science and the Social Order*. New York: Taylor & Francis.

Kakutani, M. 2018. *The Death of Truth: Notes on Falsehood in the Age of Trump*. Crown/Archetype.

Kim, Sung Ho. 2012. Max Weber. In *The Stanford Encyclopedia of Philosophy*, ed. Edward N. Zalta. Stanford: Metaphysics Research Lab, Stanford Universit.

König, Ariane. 2015. Sustainability Science. *Sustainability Hub*.

König, Ariane (ed.). 2018a. *Sustainability Science*. New York: Routledge.

König, Ariane. 2018b. Sustainability Science as a Transformative Social Learning Process. In *Sustainability science*, ed. Ariane König. New York: Routledge.

Kumar, Manasi, and Pushbam Kumar. 2008. Valuation of the Ecosystem Services: A Psycho-cultural Perspective. *Ecological Economics*.

Küppers, Bernd-Olaf. 2018. *The Computability of the World: How Far Can Science Take Us?* Springer International Publishing.

Lægreid, Per. 2017. New Public Management. In *Oxford Research Encyclopedia, Politics (politics.oxfordre.com)*, ed. Oxford University Press.

Lægreid, Per, and Tom Christensen (eds.). 2007. *Transcending New Public Management—The Transformation of Public Sector Reforms*. London: Routledge.

Larivière, Vincent, and Cassidy R. Sugimoto. 2018 (forthcoming). The Journal Impact Factor: A Brief History, Critique, and Discussion of Adverse Effects. In *Springer Handbook of Science and Technology Indicators*, ed. W. Glänzel, H.F. Moed, U. Schmoch, and M. Thelwall. Cham, Switzerland: Springer International Publishing.

Latour, Bruno. 1987. *Science in Action*. Cambridge, MA.

Lehtonen, Markku. 2015. Indicators: Tools for Informing, Monitoring or Controlling? In *The Tools of Policy Formulation—Actors, Capacities, Venues and Effects*, ed. Andrew J. Jordan and John R. Turnpenny. Cheltenham: Edward Elgar Publishing.

Lupton, Deborah. 2013. *Risk_2nd_edition*. London: Routledge.

Maggino, Filomena. 2017. *Complexity in Society: From Indicators Construction to their Synthesis*. Springer International Publishing.

Marquard, O. 2003. *Zukunft braucht Herkunft: philosophische Essays*. Reclam.

Merali, Yasmin, and David J. Snowdon. 2000. Special Editors' Note: Complexity and Knowledge Management. *Emergence: Complexity and Organization* 2: 5–6.

Merry, Sally Engle. 2011. Measuring the World—Indicators, Human Rights, and Global Governance. *Current Anthropology* 52 (Suppl 3).

Minorities, The Centre for Studies of the Holocaust and Religious. 2018. 'Innocent questions', The Centre for Studies of the Holocaust and Religious Minorities. Accessed 28.05.2018. https://publicartnorway.org/prosjekter/the-center-for-studies-of-the-holocaust-and-religious-minorities/.

Mitroff, Ian I. 2019. *Technology Run Amok—Crisis Management in the Digital Age*. Cham: Palgrave Macmillan.

Moen, Ronald D., and Clifford L. Norman. 2016. Always Applicable—Deming's System of Profound Knowledge Remains Relevant for Management and Quality Professionals Today. *Quality Progress*.

Mügge, Daniel K. 2019. The Revenge of Political Arithmetick. Economic Statistics and Political Purpose. In *Fickle Formulas*, 27. Amsterdam: University of Amsterdam.

OECD, and EuropeanCommission_JRC. 2008. *Handbook on Constructing Composite Indicators—Methodology and User Guide*. Paris: OECD.

O'Neil, C. 2016. *Weapons of Math Destruction: How Big Data Increases Inequality and Threatens Democracy*. Crown.

O'Neill, Daniel W., Andrew L. Fanning, William F. Lamb, and Julia K. Steinberger. 2018. A Good Life for All Within Planetary Boundaries. *Nature Sustainability* 1: 88–95.

PARIS 21. 2017. *Improving lives through better statistics*. http://www.paris21.org.

Patriarca, Silvana. 1996. *Numbers and Nationhood—Writing Statistics in Nineteenth-Century Italy*. Cambride: Cambride University Press.

Peruzzi, Alberto. 2017. Complexity: Between Rhetoric and Science. In *Complexity in Society: From Indicators Construction to their Synthesis*, ed. F. Maggino. Springer International Publishing.

Piketty, T. 2014. *Capital in the Twenty-First Century*. Harvard University Press.

Population Matters. 2018. Population "Factfulness"—Where Hans Rosling Goes Wrong. Population Matters, Accessed 04.09.2018. https://populationmatters.org/news/2018/04/09/population-%E2%80%9Cfactfulness%E2%80%9D-%E2%80%93-where-hans-rosling-goes-wrong.

Porter, M.E. 1980. *Competitive Strategy: Techniques for Analyzing Industries and Competitors.* Free Press.

Porter, M.E. 1990. *Competitive Advantage of Nations.* Free Press.

Porter, Theodore M. 1995. *Trust in Numbers: The Pursuit of Objectivity in Science and Public Life.* Princeton, N.J., Chichester: Princeton University Press.

Porter, Theodore M. 2004. *Karl Pearson: The Scientific Life in a Statistical Age.* Princeton, NJ; Oxford: Princeton University Press.

Porter, Theodore M. 2015. The Flight of the Indicator. In *The World of Indicators: The Making of Governmental Knowledge through Quantification (Cambridge Studies in Law and Society),* ed. R. Rottenburg, S. Merry, S. Park, and J. Mugler. Cambridge: Cambridge University Press.

Power, Michael. 1994. *The Audit Society.*

Power, Michael. 1997. From Risk Society to Audit Society. *Soziale Systeme - Zeitschrift für Soziologische Theorie* 3 (1997): 3–21.

Prigogine, I., I. Stengers, and A. Toffler. 2017. *Order Out of Chaos.* Verso Books.

Pullinger, John. 2017. Statistics are Even More Important in a 'Post-Truth' World. *The Guardian,* January 24, 2017.

Quetelet, Adolphe. 1835. *Sur L'Homme et le Développement de Ses Facultes, Ou Essai de Physique Sociale.* Paris: Bachelier, Imprimeur-Libraire.

Quine, Maria Sophia. 1990. *From Malthus to Mussolini—The Italian Eugenics Movement and Fascist Population Policy, 1890–1938.* University College London.

Radermacher, Walter. 1992. Methoden und Möglichkeiten der Qualitätsbeurteilung von statistischen Informationen aus der Fernerkundung/Methods and Possibilities of Assessing the Quality of Statistical Data of Remote Sensing. *Jahrbücher für Nationalökonomie und Statistik* 169–179.

Radermacher, Walter. 1999. Indicators, Green Accounting and Environment Statistics: Information Requirements for Sustainable Development. *International Statistical Review: A Journal of the International Statistical Institute and its Associations* 67: 339–354.

Radermacher, Walter. 2005. The Reduction of Complexity by Means of Indicators—Case Studies in the Environmental Domain. In *Statistics, Knowledge and Policy—Key Indicators to Inform Decision Making,* ed. OECD. Paris: OECD Publishing.

Radermacher, Walter. 2008. Beyond GDP—Ecosystem Services as Part of Environmental Economic Accounting? In *Workshop "Ecosystem Services—Solution for Problems or A Problem That Needs a Solution,* ed. University Kiel. Bad Salza, Germany: University Kiel.

Radermacher, Walter, and Carsten Stahmer. 1998. Material and Energy Flow Analysis in Germany: Accounting Framework, Information System, Applications. In *Environmental Accounting in Theory and Practice,* 187–211.

Radermacher, Walter J., and Anton Steurer. 2015. Do We Need Natural Capital Accounts for Measuring the Performance of Societies Towards Sustainable Development, and If So, Which Ones? *Eurostat Review on National Accounts and Macroeconomic Indicators Eurona* 2015: 7–18.

Radermacher, Walter, Roland Zieschank, Regina Hoffmann-Kroll, Jo v. Nouhuys, Dieter Schäfer, and Steffen Seibel. 1998. Entwicklung eines Indikatorensystems für den Zustand der Umwelt in der Bundesrepublik Deutschland mit Praxistest für ausgewählte Indikatoren und Bezugsräume. In *Schriftenreihe Beiträge zu den Umweltökonomischen Gesamtrechnungen.* Wiesbaden: Statistisches Bundesamt.

Randers, Jorgen, Johan Rockström, Per Espen Stoknes, Ulrich Golücke, David Collste, and Sarah Cornell. 2018. *Transformation is Feasible—How to Achieve the Sustainable Development Goals within Planetary Boundaries—A Report to the Club of Rome, for its 50 years anniversary 17 October 2018.* Stockholm: Stockholm Resilience Centre, Stockholm University, Norwegian Business School, Global Challenges Foundation.

Ravetz, Jerome. 2018. Heuristics for Sustainability Science. In *Sustainability Science,* ed. Ariane König. New York: Routledge.

Ravetz, Jerome, Paula Hild, Olivier Thunus, and Julien Bollati. 2018. Sustainability Indicators—Quality and Quantity. In *Sustainability Science*, ed. Ariane König. New York: Routledge.

Restivo, Sal (ed.). 2005. *Science, Technology, and Society*. Oxford: Oxford University Press.

Rosenblueth, Arturo, and Norbert Wiener. 1945. The Role of Models in Science. *Philosophy of Science* 12: 316–321.

Rosling, H., A.R. Rönnlund, and O. Rosling. 2018. *Factfulness: Ten Reasons We're Wrong About the World–and Why Things Are Better Than You Think*. Flatiron Books.

Rottenburg, Richard, Sally E. Merry, Sung-Joon Park Park, and Johanna Mugler (eds.). 2015. *The World of Indicators—The Making of Knowledge through Quantification*. Cambridge University Press.

Royal Statistical Society. 2014. Data Manifesto, RSS. Accessed 23.04.2018. http://www.rss.org. uk/Images/PDF/influencing-change/rss-data-manifesto-2014.pdf.

Ryan, Liz. 2014. 'If You Can't Measure It, You Can't Manage It': Not True. In *Forbes/Leadership*. Forbes.

Saetnan, Ann Rudinow, Heidi Mork Lomell, and Svein Hammer. 2011. *The Mutual Construction of Statistics and Society*. New York, NY: Routledge.

Saetnan, Ann Rudinow, Heidi Mork Lomell, and Svein Hammer. 2012. By the Very Act of Counting—The Mutual Construction of Statistics and Society. In *The Mutual Construction of Statistics and Society*, ed. Ann Rudinow Saetnan, Heidi Mork Lomell and Svein Hammer. New York: Routlegde.

Sangolt, Linda. 2010a. A Century of Quantification and "Cold Calculation." Trends in the Pursuit of Efficiency, Growth and Pre-eminence. In *Between Elightenment and Disaster—Dimensions of the Political Use of Knowledge*, ed. Linda Sangolt. Brussels: P.I.E. Peter Lang.

Sangolt, Linda. 2010b. *Between Enlightenment and Disaster: Dimensions of the Political Use of Knowledge*. Brussels: P.I.E. Peter Lang.

SDG16 Data Initiative. 2017. *SDG16 data initiative*. http://www.sdg16.org/about/.

SDSN. 2017. *Data, Indicators, and Follow-up & Review*. http://unsdsn.org/what-we-do/data-indicators-follow-up-review/.

Sébastien, Léa, Tom Bauler, and Markku Lehtonen. 2014. Can Indicators Fill the Gap Between Science and Policy? An Exploration of the (Non) Use and (Non) Influence of Indicators in EU and UK Policymaking. *Nature and Culture* 9: 316–343.

Seltzer, William. 2006. Historical Background and Some Current Concerns. In *Innocent Questions*, ed. Arnold Dreyblatt. Heidelberg: Kehrer Verlag.

Seneviratne, Amanda. 2016. Australian National Accounts: Distribution of Household Income, Consumption and Wealth. In *DGINS Conference 2016*, ed. Statistics Austria. Vienna: Statistics Austria.

Sitglitz, Joseph E. 2019. Trump's Most Worrisome Legacy. In *Project Syndicate*. Prague.

Soma, Katrine, Bertrum H. MacDonald, Catrien J.A.M. Termeer, and Paul Opdam. 2016. Introduction Article: Informational Governance and Environmental Sustainability. *Current Opinion in Environmental Sustainability* 2016: 131–139.

Stamhuis, Ida H. 2008. Statistical Thought and Practice. A Unique Approach in the History and Development of Sciences? In *The Statistical Mind in Modern Society. The Netherlands 1850–1940*, ed. I.H. Stamhuis, P.M.M. Klep and J.G.S.J. van Maarseveen. Amsterdam: aksant.

Star, Susan Leigh, and J.R. Griesemer. 1989. Institutional Ecology, 'Translations' and Boundary Objects: Amateurs and Professionals in Berkeley's Museum of Vertebrate Zoology, 1907–39. *Social Studies of Science* 19: 387–420.

Stengers, Isabelle. 2004. The Challenge of Complexity: Unfolding the Ethics of Science—In Memoriam Ilya Prigogine. *ECO* Special Double Issue 6: 92–99.

Stengers, I., M. Chase, and B. Latour. 2014. *Thinking with Whitehead: A Free and Wild Creation of Concepts*. Harvard University Press.

Stiglitz, Joseph E., Amartya Sen, and Jean-Paul Fitoussi. 2009. *Report by the Commission on the Measurement of Economic and Social Progress*.

Stiglitz, Joseph E., Jean-Paul Fitoussi, and Martine Durand (eds.). 2018b. *For Good Measure, Advancing Research on Well-being, Metrics Beyond GDP*. Paris: OECD Publishing.

Supiot, Alain. 2015a. *La Gouvernance par les nombres*. Nantes: Librairie Arthème Fayard.

Supiot, Alain. 2015b. *Le rêve de l'harmonie par le calcul*. Février: Le monde diplomatique.

Sustainable Development Solutions Network Thematic Research Network on Data and Statistics (SDSN TReNDS). 2017. Counting on the World. *Building Modern Data Systems for Sustainable Development*. In. New York: UN SDSN.

TheDemingInstitute. 2018. Seven Deadly Disease of Management. The Deminig Institute. Accessed 2.2.2018. https://deming.org/explore/seven-deadly-diseases.

Thomas, Ray. 1984. A Critique of the Rayner Review of the Government Statistical Service. *Public Administration*.

Thompson Klein, Julie. 2004. Interdisciplinarity and Complexity: An Evolving Relationship. *ECO*, Special Double Issue 6: 2–10.

Tooze, J. Adam. 2001. *Statistics and the German State, 1900–1945: The Making of Modern Economic Knowledge*. Cambridge.

TruthCommittee. 2015. Preliminary Report of the Truth Committee on Public Debt. Hellenic Parliament: Athens.

UNECE. 2014. Conference of European Statisticians Recommendations on Measuring Sustainable Development. New York and Geneva: United Nations Commission for Europe.

United Nations. 1989. *Handbook on Social Indicators*. New York: United Nations.

United Nations. 1992. Rio Declaration on Environment and Development. ed. General Assembly. Rio de Janeiro: UN.

United Nations. 2014a. Fundamental Principles of Official Statistics. New York.

United Nations. 2014b. *System of Environmental-Economic Accounting 2012—Central Framework*. New York: United Nations European Union, FAO, IMF, OECD, The World Bank.

United Nations. 2014c. *System of Environmental-Economic Accounting 2012—Experimental Ecosystem Accounting*, ed. UNSD. New York: UN, European Commission, FAO, OECD, World Bank.

United Nations. 2016. *Sustainable Development Goals*. http://www.un.org/sustainabledevelopment/sustainable-development-goals/.

United Nations. 2017a. *Framework for the Development of Environment Statistics (FDES 2013)*. New York: UN.

United Nations. 2017b. The Sustainable Development Agenda. UN. http://www.un.org/sustainabledevelopment/development-agenda/.

UNSD. 2017. The Sustainable Development Goals Report 2017. In edited by *United Nations Statistical Division*. New York: UNSD.

Van den Hove, Sybille. 2007. A Rationale for Science–Policy Interfaces. *Futures* 39.

Walton, M. 1986. *The Deming Management Method*. Perigee.

Wietog, Jutta. 2003. German Official Statistics in the Third Reich with Respect to Population Statistics. In *54th ISI World Statistics Congress*, ed. International Statistical Institute. Berlin: ISI.

Woermann, M., O. Human, and R. Preiser. 2018. General Complexity: Aphilosophical and Critical Perspective. *Emergence: Complexity and Organization* 2018: 1–17.

World Commission on Environment and Development. 1987. Our Common Future. New York: UN.

Wuppuluri, S., and F.A. Doria. 2018. *The Map and the Territory: Exploring the Foundations of Science, Thought and Reality*. Springer International Publishing.

Zak, Paul. 2013. Measurement Myopia. In *Drucker Institute*, ed. Drucker Institute. Drucker Institute.

Zamora, Daniel, and Michael C. Behrent. 2014. *Foucault and Neoliberalism*. Cambridge.

Chapter 4
Official Statistics 4.0: The Era of Digitisation and Globalisation

What the future, even the very near future, has in store for us is, of course, not easily predictable. Certainly, foresight does not lie in the very nature of statisticians, who usually look in the rear-view mirror. Nevertheless, there are some trends or megatrends, the effects of which are not yet known in detail, to which one will undoubtedly have to adapt. Above all, because official statistics have the characteristics of an ocean liner whose course and speed can only be manoeuvred slowly, all trends must be interpreted in a forward-looking manner. If official statistics are to be sustained in their current position in five years' time, then the necessary strategy must be established now. A simple continuation of the previous way of doing things, but including some 'softer' changes, is therefore not an option, even if this smooth manner of adaptation has been successful in the past. The following chapter addresses the two megatrends of digitisation and globalisation. Obviously, it is not the purpose of this work to deal with their methodological, conceptual, or technical aspects in this regard. Rather, it is about the consequences in terms of the statistical policy due to the changed statistical environment and conditions.

4.1 Facts for Future—Which Future? Which Evidence?

For a strategic planning of the medium-term and long-term developments of the statistical programme (including its products and processes), foresighted, scientifically founded reports are of great importance. After all, it usually takes years for the statistical ocean liner to complete an appropriate manoeuvre and adjust its course. At the end of 2018 and the beginning of 2019, there was a whole series of such reports relevant to statistical planning:

- Transformation is Feasible—How to Achieve the Sustainable Development Goals within Planetary Boundaries (Randers et al. 2018)
- For Good Measure—Advancing Research on Well-being Metrics Beyond GDP (Stiglitz et al. 2018b)

© Springer Nature Switzerland AG 2020
W. J. Radermacher, *Official Statistics 4.0*,
https://doi.org/10.1007/978-3-030-31492-7_4

- Beyond GDP—Measuring What Counts for Economic and Social Performance (Stiglitz et al. 2018a)
- Sustainable Equality—Well-Being for Everyone in a Sustainable Europe (Independent Commission for Sustainable Equality 2018)
- Work for a Brighter Future (Global Commission on the Future of Work 2019)
- The Global Risks Report 2019 (World Economic Forum 2019)
- Global Trends to 2013—Challenges and Choices for Europe (European Strategy and Policy Analysis System 2019)
- Global Assessment Report on Biodiversity and Ecosystem Services (Intergovernmental Science-Policy Platform on Biodiversity and Ecosystem Services 2019).

Although these reports cover a very wide range of topics and focus on different aspects, there is a high degree of agreement on future (global) risks. For example, the representatives of business surveyed by the World Economic Forum rate the following risks as the most serious in terms of likelihood of occurrence and impact (in decreasing order): *"extreme weather events, failure of climate-change mitigation, natural disasters, cyber-attacks, water crises, biodiversity loss and ecosystem collapse, man-made environmental disasters, large-scale involuntary migration, interstate conflict and failure of regional and global governance"* (World Economic Forum 2019, p. 5). Moreover, while all these reports place great emphasis on the need for sound information to manage all these risks and the social upheaval they entail, it is imperative that all attention and prioritisation of long-term planning follow this assessment. Although a movement has been set in motion with the Sustainable Development Indicators, the question remains whether this individual activity is sufficient or whether a more fundamental revision of the statistical programmes with a reprioritisation in favour of these fields would not be necessary.

4.2   Rapid and Radical Changes—The New Environment for Official Statistics

4.2.1   Three Revolutions in the Digital Age

The digital age is not just a gradual evolution of previous phases of information and communication technology. Rather, a profound change is taking place in society, which fundamentally changes personal behaviour in everyday life, and leads to completely new mixtures of risks and opportunities, of winners and losers and of consumers and producers concerning data or information. It is spoken of as a data revolution,[1] to clarify the extent of the current structural change; however, technological changes do not happen in a vacuum, but are continually influenced by, and

---

[1] See, for example, the data revolution group, established by the UN (http://www.undatarevolution.org/).

influence themselves social and political conditions, both of which are witnessing major changes. Overall, the following three developments are of prime importance for the future of official statistics:

First: Zettabytes and yottabytes

The era of the data revolution has started, significantly changing the picture with regard to both the production and consumption of data. On one hand, the availability of enormous amounts of data gives the statistical business a completely new push in a direction that is not yet sufficiently understood—although there is growing awareness for the synergies and potentials of close cooperation between statistics and other disciplines of data science (Cao 2017a, b).

In recent years, the quantity of digital data created, stored and processed in the world has grown exponentially. Every second, governments and public institutions, private businesses, associations and even citizens generate series of digital imprints which, given their size, are referred to as 'Big Data'. The wealth of information is such that it has been necessary to invent new units of measurement, such as zettabytes or yottabytes, and sophisticated storage devices purely to deal with the constant flow of data. The world can now be considered as an immense source of data. Broad consensus reigns with regard to the wonderful opportunities which 'Big Data' can bring in relation to the statistics acquired from traditional sources such as surveys and administrative records. These opportunities include:

- Much faster and more frequent dissemination of data.
- Responses of greater relevance to the specific requests of users, since the gaps left by traditional statistical production are filled.
- Refinement of existing measures, development of new indicators, and the opening of new avenues for research.
- A substantial reduction in the burden on persons or businesses approached and a decrease in the non-response rate.
- Last, but not least, access to 'Big Data' could considerably reduce the costs of statistical production, at a time of severe cutbacks in resources and expenditure.

However, 'Big Data' also threatens a number of challenges:

- These data are not the result of a statistical production process designed in accordance with standard practice.
- They do not fit current methodologies, classifications and definitions, and are therefore difficult to harmonise and convey in the existing statistical structures.
- Complex aggregates, such as the gross domestic product (GDP) or the Consumer Price Index (CPI) aim at measuring macro-economic indicators (Lehtonen 2015) for the nation as a whole; their substitution by Big Data sources seems to be out of reach.
- In addition to this, 'Big Data' raises many major legal issues: security and confidentiality of data, respect for private life, data ownership, etc.

All of the above mean that, at least for now, 'Big Data' can only be used to a limited degree to supplement, rather than replace, sources of traditional data in certain statistical fields.

Second: Evidence and decisions

On the other hand, the demand for 'evidence-based decision-making', (new public) management, and other applications of a neoliberal governance model (Davies 2016) create a powerful driving force on the demand side of statistics. It can be recognised as an *"ingredient of rationality"* (Peruzzi 2017, p. 4) to take into account the consequences of a decision. It is a long way from this 'Enlightenment' viewpoint to a form of governance in which the availability of evidence is considered a prerequisite for any decision. *"During the past hundred years or so, political governance underwent a massive "quantitative turn." This quantitative turn is here understood as systematic effort to delineate and measure the objects and results of governance quantitatively for the purpose of demonstrating competitive edge and superiority at the individual and/or collective level."* (Sangolt 2010a) Now, from a statistical point of view, it seems almost desirable that this quantitative turn has led to a greater demand and supply of statistics, if there were not a number of side effects, which could endanger the quality of statistical information or could even be a threat to official statistics (see Chap. 3). If it is true that *"measurement is a religion in the business world"* (Ryan 2014), this religion not only has a significant impact on the behaviour of managers, civil servants, and public and private sector employees, but rather, it also creates a hunger for data that is not matched by the appetite for good quality. In such an 'audit society', there is a great danger that the existence of data and information is assumed to be normal. That these informational products must be produced, that they can have indifferent quality, and that producing them costs time and money, is quickly pushed into the background when it comes down to having any data whatsoever available. Paradoxically, this same information society complains that the burden of statistics is too high. In all of this, it becomes clear that significant risks to statistics can arise because expectations concerning their quantity are too high, while those concerning their quality are too low. It is difficult to sustain a high-quality profile of products in a fast food culture.

All these trends, which have emerged in recent decades, are being accelerated by new technologies. Decisions which were 'augmented' by the use of evidence might now become 'automated'. The Internet of Things (IoT) (De Clerck 2017), artificial intelligence (AI) and the growing importance of algorithms are posing new questions in areas other than technological ones[2]: *"Society must grapple with the ways in which algorithms are being used in government and industry so that adequate mechanisms for accountability are built into these systems. There is much research still to be done to understand the appropriate dimensions and modalities for algorithmic transparency, how to enable interactive modelling, how journalism should evolve, and how to make machine learning and software engineering sensitive to, and effective in, addressing these issues"* (Diakopoulos 2015).

Third: Facts and alternatives

"It is so comfortable to be a minor. If I have a book that thinks for me, […] then I have no need to exert myself." (Braungardt 2018) Immanuel Kant wrote this in 1784

---

[2]See the report 'For a meaningful artificial intelligence—Towards a French and European Strategy' (Villani 2018).

to rouse his contemporaries into thinking for themselves. But what does the legacy of the Enlightenment look like today? What is the contribution of statistics? Statistical history may well be rooted in our cultural, political and social backgrounds, but as these change—and become more international—how do the resulting tensions play out?

The Oxford Dictionaries, and the Society for the German Language, have both chosen 'post-truth' (OxfordDictionairies 2016) [or in German 'post-faktisch' (GfdS 2016)] as Word of the Year 2016.

> In recent years, as the European Union has expanded, there has been growing mistrust on the part of citizens towards institutions considered to be aloof, engaged in laying down rules perceived as insensitive to individual peoples, if not downright harmful ... As a result, the great ideas which once inspired Europe seem to have lost their attraction, only to be replaced by the bureaucratic technicalities of its institutions. Pope Francis (2014)

This short selection of quotes could easily be expanded by referring to the ongoing debate in the follow-up to the unexpected political disruptions in 2016. Information and facts are not neutral. Just as other manufactured products, they open manifold possibilities of 'dual' use and of risks which must be anticipated by responsible information producers in their policies and production processes. One of the key questions that, again, has to be asked, is related to the role that sciences have played in the past, and in how far this role needs to be critically assessed and revised (Benessia et al. 2016; Beck 1998).

While uncertainties and risks are constantly growing in the eyes of citizens, and while the impact of globalisation becomes more and more visible, it appears as if people have had enough of experts (Mance 2016). It also appears as if 'post-truth-politics' would gain credibility and support, opening opportunities for populist and nationalist activists of all kinds. The trust of the population in their governments, and in official institutions in general, is rapidly decaying, and this lack of trust is naturally extended to the producers of official statistics.

Citizens ask themselves what use statistical indicators serve, and for whose benefit. Knowledge is power. Is statistical evidence used to stimulate political dialogue (opening up), to shorten it, or, in the worst case, to suppress it (closing down) (Lehtonen 2015)? Depending on how these questions are answered, statistics will win or lose citizens' trust. The closeness of official statistics to politics and their embeddedness in public administration can have both positive and negative consequences, depending on the perception of their use in political decision making and their professional independence.

In this context, a profound epistemological shift is needed since complexity and irreversibility undermine the idea that science (and statistics) can provide single, objective and exhaustive answers. In the late modernity of risk societies (Beck 1998), there is the epistemic and methodological necessity to empower people—citizens and policy makers—with the appropriate insight, to enable them to make the best possible decisions for achieving sustainability and pursuing resilience in a complex world: *"The tools for thought of the Enlightenment no longer suffice for mastering the challenges of the present. The course European societies are taking can be compared to the exploration journeys of bygone days. Maps, which ought to provide orientation and security, seem to have lost their value. We are journeying into the uncertain and have yet to discover new paths and routes in many areas."* (European Alpbach Forum 2016)

4.2.2 Globalisation: National Statistics Under Pressure

If we follow Desrosières, then the marriage between the nation state and statistics was a result of the enlightenment (Desrosières 1998). The preparation and establishment of nation states (for example, in the Risorgimento in the first half of the nineteenth century in Italy) required reliable statistics, as well as political decisions in the states. Similarly, the emergence of international (e.g. UN) and supranational (EU) institutions has been accompanied by the development of statistical infrastructure.

It is not surprising, therefore, that at a time when fundamental questions regarding the role of the nation state are raised and discussed (Hale et al. 2013; Dasgupta 2018), statistics are also affected. The nation state, and its ability to solve pressing problems, comes under pressure from two directions: on the one hand, global problems (global value chains, migration, global environmental problems, etc.) require stronger international cooperation, which is partly accompanied by shifts in power and influence. On the other hand, at least in the larger states, individual regions claim greater freedom of choice and responsibility for political fields.

Official statistics must adapt to this situation, calling their usual paradigms into question to provide new, revised, answers:

- First, it is about finding relevant and measurable indicators of globalisation which, by definition, are not coincident with the borders of nation states. Equally, the set of indicators that are in use for the measurement of progress of societies (such as GDP, employment, productivity) need to be complemented by progress indicators tailored primarily for international purposes.
- Second, the question arises whether these indicators can be identified, and produced, using adequate methods as long as their production takes place within national statistical systems and paradigms.
- Third, it is about finding new ways of cooperation across borders, and the exchange of data and knowledge between the actors of official statistics.
- Finally, it will also be about using new technical possibilities and data sources (geo-coded data, GIS, remote sensing) to produce sub-national (regional, local) statistical results of high quality.

4.2.3 Official Statistics 4.0: Answers to a Dramatically Changing Environment

In the preceding chapters, the evolution and history of official statistics over the last two hundred years have been used to explain which forces (science, statistics, society) are drivers of development, and how they interact. The status, called **Official Statistics 3.0**,[3] has allowed official statistics to deal with all the risks and environmental factors of technology, data and politics, known up to the beginning of this century.

The evolution of digitisation and globalisation changes the working conditions, tasks and ecosystems of official statistics. The interplay of, and feedback loops between, all driving forces create great uncertainty and dynamism. As a result, we are now witnessing the appearance of a number of new challenges requiring innovative answers and solutions [Martín-Guzman 2018 (forthcoming)].

The continuous, bottom-up improvement of processes, technologies and data sources that has characterised the last decades of official statistics is not enough in such an era of dramatic changes. The completely new, competitive situation requires official statistics to provide innovative strategic answers that go beyond traditional statistical methods and technologies. The core of this will be to maintain (or, if already lost, to win back) trust in official statistics, both as an institution and as an information infrastructure, in the face of scepticism towards politics and state institutions.

In the age of Big Data, AI and algorithms, a need exists for ethical guidance and legal frameworks under new conditions: *"In the world being opened up by data science and artificial intelligence, a version of the basic principle of the partnership between humans and technology still holds. Be guided by the technology, not ruled by it"* (Lohr 2016). What might facilitate the perceived new search for orientation and balance is the stock of ethical and governance principles that is available, emerging from two hundred years of history in official statistics.

For some years, especially in the field of environmental data, a new form of cooperation between science and citizens, called 'citizen science', is developing.[4] Citizen science projects actively involve citizens (as contributors, collaborators, etc.) in scientific endeavour that generates new knowledge or understanding.[5] Although citizen science is still relatively young, it hits the point, which is becoming increasingly important for official statistics. The past distinction between producers of data and consumers makes less and less sense. Consequently, the question arises of how to actively involve citizens in the production of statistics throughout the entire production chain, from design to communication. In the past, citizens (as well as companies and many other partners of statistics)[6] were either passive respondents in surveys and/or simply consumers of ready-made statistical information. The answer to this

---

[3] See Chap. 2.
[4] See, for example, Haklay *"Citizen Science and Policy: A European Perspective"* (Haklay 2015).
[5] See ECSA (2016).
[6] See Soma et al. (2016a).

question is anything but trivial. At its core are the same problems and difficulties as the issue of using Big Data for official statistics in general: control of procedures, quality assurance, interpretability of information and neutrality/impartiality.

In a later section of this chapter, we will go into more detail about these questions.

4.2.4 Launching a New, Scientific Debate

It seems to be both necessary and urgent to launch a scientific debate in professional communities and initiate a period of reflection. Research is needed on the role of official statistics in society, and vice versa, making use of scientific concepts such as 'co-production' (Jasanoff 2004a) or "*gouvernementalité*" (Foucault 1991).[7] Scientific research and development are essential to the quality of measurements and their results, whether they are based on statistical survey methodologies, or driven by data science concepts. Apparently, this relates in the first place to the relevant technical disciplines (Hand 2004). However, this should be supplemented by going beyond pure methodologies, by taking on board aspects from other fields, such as sociology, historical or legal disciplines. There are many different strands of science contributing research on processes of quantification and the impact of quantification within social contexts (Diaz-Bone and Didier 2016). Those scientific inputs should address questions and issues such as:

- phases in the history of official statistics having the potential to explain the interaction between knowledge generation and society; the making of states; statistics under authoritarian, liberal and neoliberal regimes
- official statistics as part of a knowledge base for life
- historical, cultural and governance systems of countries; differences between statistical authorities, and their performance across the globe compared to in Europe; international/supranational governance in statistics
- creation of knowledge; measurement in science and practice; limits of measurement; facts and (science) fiction; statistics and theories, such as economic theory, epistemology and falsification/verification of theories
- use, misuse and abuse of evidence; the power of knowledge and how to share it; relationship to conceptual frames in politics
- public value in the context of public administration; participation of citizens via effective and efficient mechanisms
- (new) Enlightenment; knowledge for the empowerment of citizens; citizen science; statistical literacy; education; participation in decision making; fostering the democratic process
- communication of data and metadata, and quality for users with unequal pre-knowledge and statistical literacy
- framing of indicators as a co-design process that activates the interest of civil society

---

[7]See Chap. 3.

- co-production of statistics; turning users of statistics into co-producers ('pro-sumers')
- quality of information, institutions, products and processes; how to decide on conventions about methodologies and programmes of work; quality assurance
- professional ethics (for individuals) and good governance (for institutions)
- professional profiles: survey methodologist, data scientist, accountant, data architect, social science engineer, etc.

Some of these issues are discussed in more detail in the next part of this chapter, while others require further investigation and wider participation.

4.2.5 Principles of Official Statistics in the Era of Digitisation

Before going into more detail about some of the important aspects for the future of official statistics, the baselines are condensed into a few guiding principles.[8]

High-quality, official statistics strengthen democracy by allowing citizens access to key information that enhances accountability. Access to robust statistics is a fundamental right that facilitates choices and decisions based on valid information. Without statistics, there cannot be a well-grounded, modern, or participatory democracy. Statistics is key for people empowerment.

Official statistics are the cornerstone of public open data; they are the basis of open government. For example, on the EU Open Data Portal, the Eurostat statistical database accounts for the bulk of data on offer. Enhancing access to statistics in open formats enables the free use of data, its interoperability and its consumption in integrated modalities. As a result, open statistics allow citizens to make sense of complex phenomena and help in their interpretation across borders and without limits. As such, open data and open statistics are a key driver of free dialogue in open societies.

Statistical literacy is critical in ensuring that individuals benefit from the power of statistics and can benefit from open access to statistical information and its associated services. Data literacy ('datacy') is not limited to knowledge of basic statistical information: it entails knowing about the limits of statistics and their use/misuse. The ability to understand statistics, and how they are produced, is a fundamental skill for each individual. 'Datacy' is a key enabler for citizens.

Data for statistical services is worthless unless statistical methods are in place to ensure quality. In the digital ecosystem, where data is abundant and a commodity, the value of information is increasingly based on algorithms that generate tailored insights for users. The future is (trusted) smart statistics.[9]

---

[8]The principles were presented in the Conference of European Statistical Stakeholders CESS 2016 in Budapest (Radermacher and Baldacci 2016).

[9]See the "Bucharest Memorandum" (European Statistical System Committee 2018) and in particular the paper on a reference architecture for smart statistics (Ricciato et al. 2018).

On the whole, the general public is distanced from official statistics and valuable statistical information. Hence, a bridge must be built between experts and laypeople to overcome this distance and to foster understanding. Providing better information to users and non-users, and being able to counter misjudgements and prejudices with facts, is probably the part of the statistical mission that has the greatest added social value. That mission is about education and providing information that is orientated towards the layperson. However, it should also be about co-design and co-production, with the overall aim of involving the public in the generation of statistical results.

As statistical information is increasingly used for policy decisions, statisticians need to investigate how their services are used, not used or misused. They should also examine the ethical implications and the impact of evidence use on the policy cycle. More influence means more responsibilities.

Box 4.1 Guiding principles of Official Statistics 4.0

- **Statistics is key for people empowerment**: Statisticians should be aware of data's power to provide information and, hence, knowledge.
- **Open data is fundamental for open societies**: Statisticians should ensure open and transparent access to data and metadata, and monitor their actual use for information and knowledge.
- **'Datacy' is a key enabler for citizens**: Statisticians should promote data literacy in society at large, and regularly monitor the levels of understanding.
- **The future is smart statistics**: Statisticians should continue to invest in methods, algorithms and a business architecture that enhance the quality of data for statistical services tailored to users' needs.
- **Users participate in the design, production and communication of statistics**: Statisticians should foster a greater involvement of civil society in all stages and processes of statistical production.
- **More influence means more responsibilities**: It is the duty of statisticians to explore the link between statistics, science and society and to lead intellectual reflections on the possible risk of over-reliance on data-centrism.

4.3 Globalisation—Reviewing the National Statistics Paradigm

Globalisation is the second driver behind rapidly changing requirements for, and working conditions of, official statistics. The manual of the European System of Accounts—ESA 2010 introduces globalisation in the following way: *"The increasingly global nature of economic activity has increased international trade in all its forms, and increased the challenges to countries of recording their domestic economies in the national accounts. Globalisation is the dynamic and multidimensional process whereby national resources become more internationally mobile, while national economies become increasingly interdependent. … All of these increasingly common aspects of globalisation make the capture and accurate measurement of cross-border flows a growing challenge for national statisticians. Even with a comprehensive and robust collection and measurement system for the entries in the rest of the world sector (and thus also in the international accounts found in the balance of payments), globalisation will increase the need for extra efforts to maintain the quality of national accounts for all economies and groupings of economies"* (Eurostat 2013b, p. 3).

The Review of UK Economic Statistics by Charles Bean highlights the intersection of digitisation and globalisation as a driving force of change: *"Measuring the economy has become even more challenging in recent times, in part as a consequence of the digital revolution. Quality improvements and product innovation have been especially rapid in the field of information technology. Not only are such quality improvements themselves difficult to measure, but they have also made possible completely new ways of exchanging and providing services. …Moreover, while measuring physical capital – machinery and structures – is hard enough, in the modern economy, intangible and unobservable knowledge-based assets have become increasingly important. Finally, businesses such as Google operate across national boundaries in ways that can render it difficult to allocate value added to particular countries in a meaningful fashion. Measuring the economy has never been harder"* (Bean 2016, p. 3).

These two statements only prove the importance of globalisation and digitisation for the economy and for economic statistics. However, there are also far-reaching changes in societies (migration, integration) and in the environment (climate, biosphere, global boundaries), which correspond to new requests for social and environmental indicators.

Charles Bean's report highlights the key challenges arising from these changed conditions and makes specific recommendations concerning the statistical programme (e.g. better coverage of the service industry, regional statistics). It also requests greater agility in statistics and adaptability to meet user needs, better use of existing data and technologies, and finally stresses the importance of statistical governance.

While many of the thoughts below have a lot of overlap with the Bean Report, they are generated with a different perspective in mind, and so are complementary. The question here is where the paradigm that is fundamental for official statistics at **national** level reaches its limits, and needs to be reflected and complemented by European or international statistics.

The National Statistics paradigm, described in Sect. 2.4, is recalled. The three dimensions of national official statistics are:

- Temporal dimension: a fixed time period (often a year, but sometimes also a quarter or a month) or a fixed date (for example, for the population census).

- Spatial dimension: a country (political delimitation), or a region (province, local unit according to administrative delimitation according to the Nomenclature of Territorial Units for Statistics).
- Measurement object: resident population and their activities; methodologies (variables, classifications, sampling schemes, etc.) designed to address national needs and priorities.

The gross domestic product—as the name implies—represents a section of the world economy capturing the activities (i.e. flows) in one country, in one period as accurately as possible. The underlying logic of economic statistics assumes that the essential production of goods physically takes place in the respective country. Cross-border activities (e.g. international trade and foreign investments), and cross-period activities (e.g. depreciation of assets), are amalgamated into the GDP of one specific country, in one year with the best possible accuracy. Quantifying the international aspect of economic cooperation (in particular for services) has itself traditionally been a low priority task. Similarly, the balancing of stocks (caused by depreciation and investments) has traditionally played a subordinate role in the concept of GDP, which focuses on the quantification of flows. Both features are now critically questioned: it is demanded that stocks are better quantified, and better statistics of global economic cooperation are required.

The report on global value chains by Timothy Sturgeon highlights: "*International trade and foreign direct investment (FDI) have long been important features of the world economy, and both have grown steadily since the end of World War Two. …Today the picture has grown more complex, with multilayered international sourcing networks and new technology-enabled business models that better integrate and accelerate cross-border economic activity*" (Sturgeon 2013, p. 1). With a sense of urgency, Sturgeon requests a far reaching programme for the statistical treatment of globalisation: "*The greater scale, complexity, and transformational potential of economic globalization demand that we ask more from our economic statistics: ways to systematically differentiate arms-length trade from intra-group trade and external international sourcing, ways to track services trade in more detail, ways to determine the real location of value added, and ways to differentiate globally-engaged from non-globally-engaged enterprises so the performance of these very different segments of national economies can be tracked in terms of profits, innovation, employment, and wages paid*" (Sturgeon 2013, p. 6).

In particular, the so-called Irish case—when the Central Statistical Office (CSO) of Ireland published a level shift for its GDP (caused by relocation of the seat of a multinational company), significantly revising the growth rates for 2015 upwards to 26.3 ppt[10]—has clarified the urgency and brought the size of the challenge into sharp relief.

After this case, the international statistical community has intensified the search for new solutions to this problem, which could include new indicators for improving insight to national economies. Further, it might become necessary to modify or extend

---

[10]See ESRG (2016), Stapel-Weber and Verrinder (2016).

statistical standards[11] to avoid the distortions caused by multinational enterprises that arrange cross-border business practices for the purposes of tax avoidance.[12] Finally, one might have to arrange new international statistical components, particularly in Europe. A deeper and more holistic review of methodologies (National Accounts and business statistics), seems to be necessary,[13] in which new statistical infrastructural elements and new international cooperation models are agreed between statistical compilers.

In recent years, international cooperation has successfully led to new indicators that can quantify key aspects of globalisation. Above all, these include the interlinked input–output tables[14] of individual countries, and the calculation of trade-in-value-added (TIVA).[15] These input–output tables prepare the ground at a global level for an economic analysis of globalisation, and for environmental indicators that measure pollutants or other environmental pressures. Among these measures are the 'Carbon Footprint'[16] or the 'Total Material Consumption (TMC)', which differs from the 'Domestic Material Consumption (DMC)' which excludes the environmental impacts of imports or exports.[17]

Registers of business statistics allow the development of recent years to be traced and an outlook for the near future to be derived. After the establishment of individual national business registers in the 1980s, a European standard format was agreed,[18] then a EuroGroups Register (EGR)[19] was created, which is currently being further developed into the European System of interoperable Business Registers (ESBRs).[20] Future tasks of this working area in official statistics will contain the so-called profiling of large cases and multinational companies.[21]

It makes little sense that each country's statistical office builds a new team for profiling in isolation, only the national segments of multinational companies. Profiling requires that the statistical data correctly monitors the overarching multinational in its entirety, as well as each of its national sub-elements according to the usual statistical

---

[11] See, for example, the Guide to Measuring Global Production (https://www.unece.org/index.php?id=42106).

[12] See Moulton and Ven (2018).

[13] See Stapel-Weber et al. (2018).

[14] See https://www.oecd.org/sti/ind/inter-country-input-output-tables.htm or http://www.wiod.org/home.

[15] See https://unstats.un.org/unsd/trade/globalforum/trade-value-added.asp or https://www.oecd.org/sti/ind/measuring-trade-in-value-added.htm.

[16] See http://ec.europa.eu/eurostat/statistics-explained/index.php/Greenhouse_gas_emission_statistics_-_carbon_footprints.

[17] For both indicators see the definition here https://www.un.org/esa/sustdev/natlinfo/indicators/methodology_sheets/consumption_production/domestic_material_consumption.pdf.

[18] See http://ec.europa.eu/eurostat/statistics-explained/index.php/Business_registers.

[19] See http://ec.europa.eu/eurostat/web/structural-business-statistics/structural-business-statistics/eurogroups-register.

[20] See http://ec.europa.eu/eurostat/web/ess/esbr.

[21] See https://ec.europa.eu/eurostat/cros/content/profiling-esbrs_en.

methods. Double counting is just as inadmissible as statistical gaps. Such a consistent statistical profile of a multinational company can only be achieved through close cooperation and data exchange, preferably combined with a strong work-sharing ethic.

However, this division of labour affects the affiliation of the statistical offices to the public administrations of each individual country. In turn, this leads to considerable problems sharing and transferring responsibilities and in exchanging data. This difficulty is demonstrated by the example of trade statistics between the European Member States. Following the introduction of the European Single Market, statistical evaluations of border controls (and thus customs declarations for the movement of goods) were stopped. The collection procedure Intrastat was introduced instead, the complete opposite of the trend, common today, of replacing surveys with administrative data. To fully satisfy all the data needs of the past, all exports, as well as all imports, were collected. Theoretically, the sum of all European exports would have to be identical to all imports, so the imports of one country could, in principle, be calculated by summing up the exports of all its neighbours. Nonetheless, a two-flow concept was established that has been in place for more than two decades, despite involving extremely high costs and burdens. One of the main reasons for keeping this procedure from a national point of view is that the so-called mirror differences (i.e. the discrepancy between collected and calculated imports) are solved by systematically favouring data collected nationally. As a consequence, while the Intrastat process produces 28 results (each being of the best quality from a national point of view), they are all contradictory and do not result in a plausible, consolidated figure at European level.

After years of difficult negotiation, a compromise has been found based on the exchange of confidential export data between the statistical offices of all Member States. For this exchange to work, strict principles have been agreed to ensure the protection of this sensitive data. It remains to be seen how this new and creative process will work in practice.

After all, this completely new form of cooperation between national statistical offices can be regarded as a successful first step into the future. A next important step could be taken with a work-sharing procedure for profiling.

Unfortunately, there are currently also signs that the national statistics paradigm is likely to be reinforced. For example, the overall architecture and governance of the Sustainable Development Goals and Indicators programme follows a political doctrine, according to which the UN Member States each take the lead in implementing both the political strategy and the corresponding indicators. One might ask whether this will result in a fragmented landscape of national indicator sets, rather than capturing global aspects of sustainability.

A very special field of work is that of public finance statistics in Europe. This is about indicators (debt and deficit) with extraordinary political significance, which does not allow any scope for national variants in the production of statistics. With such close integration and interdependence, it would be worth considering a different statistical system with a more centralised structure and governance to replace the current decentralised system of national and European institutions (Georgiou 2018).

However, such considerations have so far had little chance of being realised, although they could offer major advantages in terms of effectiveness and efficiency.

4.4 Bridging the Gap—Communication 4.0

"Bridging the Gap between Citizens and Official Statistics" is the title of Corine Eyraud's contribution (Eyraud 2018) to the Power from Statistics Conference in 2017. *"Citizens' growing suspicion vis-à-vis the official statistics, suspicion which would be in line with our 'post-truth' and anti-intellectualist era"* is her starting point for a rather new entrée to the question of communicating statistics. *"It can be acknowledged that statistics have regularly been used by politicians or managers (from public and private sectors) to mislead people, to justify political and economic decisions pretending them to be evidence- based, or to make them so difficult to understand that non-expert people will not be able to question the choices and decisions which are made. Hence, statistics have been part of the system of domination. The first thing to do to bridge the gap between citizens and statistics will be to stop using them in that way and for that kind of purpose"* (Eyraud 2018, p. 103).

How can we best bridge the gap between the public (the 'citizen') and statisticians? Is it enough to focus on improving the communication of statistical results? Is the problem to be solved purely one of language? Or do we need to start further upstream in the sequence of processes of measurement/quantification[22] described in Chap. 2 (design, production, communication), and address the production of statistics, as well as the process of knowledge creation by users? Does the communication of the future perhaps require more participation? If so, who should participate and how should this be done in practice?

In the following, some approaches will be pursued that focus primarily on mainstreaming users and their interests throughout the production process. Most importantly, however, it is a question of fostering a greater involvement of civil society; that is to say, the general public are, on the whole, somewhat distanced from official statistics and valuable statistical information, so a bridge must be built in order to overcome that distance. Providing better information to users and non-users, and being able to counter their misjudgements and prejudices with facts, is probably

---

[22] 'However, till very recently, very few studies have questioned the figures they used, as if these figures were simply measuring a pre-existing reality. To prevent this "realist epistemology", Alain Desrosières, who is the founder of a new way of thinking about statistics, proposed to talk not about "measurement" but about "quantifying process": "The use of the verb 'to measure' is misleading because it overshadows the conventions at the foundation of quantification. The verb 'quantify', in its transitive form ('make into a number', 'put a figure on', 'numericize'), presupposes that a series of prior equivalence conventions has been developed and made explicit [...]. Measurement, strictly understood, comes afterwards [...]. From this viewpoint, quantification splits into two moments: convention and measurement."' (Eyraud 2018, p. 103).

the part of the statistical mission that has the greatest added social value.[23] According to the legacy of Hans Rosling,[24] that mission is about education and providing information that is orientated towards the layperson. However, it should also be about co-design and co-production, through which the participation of the public in statistical results should be the aim.

Of course, the involvement of users and their interests has always played a significant role in official statistics. During the development and revision of both the statistics programme and of individual statistics, user advisory councils are consulted, scientific colloquia are organised, and, finally, legal decision-making processes are followed. The critical aspect here is that it is essentially a very narrow selection of experts and stakeholders who are involved in such consultation processes.

The dissemination of statistical information has undergone a complete transformation in recent years. This has started with the fact that the term 'dissemination' is now largely shunned and has been replaced with 'communication'. In place of a publication programme producing a single flagship *Statistical Yearbook*, a series of individual, specialised and very wide-ranging (printed or online) books has emerged. These are geared towards online media and have social networks as integrated distribution channels. Statistical offices commonly have an Internet presence and websites prepared for diverse user groups as standard. Interactive communication tools and mobile applications facilitate access, even for the layperson.

Nevertheless, there is more to do. With reference to the still relatively young discipline of 'citizen science',[25] we need to understand the circumstances that have led to the mistrust of the elite in Western society, and the way that statistics are (or are at least perceived to be) an instrument of both the political/administrative elite and the scientific elite. William Davies' analysis (Davies 2017) could be taken as a starting point for reflection on the challenges and opportunities brought by this rapidly changing environment. A few of his observations, all of which add up to a general mistrust of official statistics, are as follows:

- Misunderstanding the real meaning of indicators by a society with a poor level of statistical literacy can create:

  - incorrect opinions
  - which may mislead voters or
  - compel politicians to take non-optimal measures.

- Advocating the objectivity and expertise of technocrats as a better choice than the regime of demagogues/politicians is associated with the following risks:

---

[23] See for example, Roser (2018): '*Most of us are wrong about how the world has changed (especially those who are pessimistic about the future)*'.

[24] Hans Rosling was a physician and statistician who, with his passion and his gift for explanation, managed to portray statistics completely new ways and use completely new dimensions of communication; he died in February 2017 (https://www.theguardian.com/global-development/2017/feb/07/hans-rosling-obituary and https://www.gapminder.org/).

[25] See Haklay (2015): *Citizen Science and Policy: A European Perspective* (Haklay 2015).

- high-level aggregated artefacts (e.g. GDP) may be too abstract in their design and meaning for the average layperson
- ex ante/top-down classifications are out of touch with the identities of individuals
- national policies are too distant from individuals and their private spheres
- in our era of Big Data, data-driven logic (the inductive search for messages in the data) has replaced statistical logic (top-down design of classifications and variables to be surveyed)
- social network bubbles undermine the existence of facts.

The public's mistrust of elites and technocrats, and their sympathy for demagogues and populists, may not seem rational.[26] Nonetheless, it is a real, international and serious phenomenon of our current time.

What are the consequences for official statistics, if confidence in public institutions is generally shrinking, if the authority of the state and its representatives is questioned, and if facts are no longer seen as being without alternative?

The circular flow of statistical processes (design, production, communication, use; see Fig. 4.1) needs to be reviewed, wherever possible, aiming to bring on board both stakeholders and civil society: in their design (e.g. the early involvement of the public regarding new indicators and data platforms during their planning stages; human-centred co-design), in their production (e.g. crowd-sourcing of data; co-production) and in their communication (which should be interactive, open, accessible, etc.) and in their use (by collecting evidence through market research of the use/misuse/non-use of indicators, by creating user-specific feedback loops, and by improving statistical literacy).

First and foremost in the future-orientated involvement of users is to remove the mental separation between the producers and the consumers of statistics. To do this, it is necessary to anchor the goal of involving civil society as deeply as possible in the production process. The most important thing to do first is make people aware of the importance and consequences of statistics and numbers in their own lives and societies. A more fare reaching objective would aim at consumers becoming co-producers ('prosumers'); stakeholders becoming shareholders. Similar to the introduction of the primacy of existing data over new surveys in the 2000s, change needs to be achieved in a well-rehearsed and conservative-thinking sphere; patterns must be maintained by defining strategic goals. The strategic goal here is to intensify the partnership between civil society and statistics in all the stages of the latter: in the scientific and design phase, during production, and—most importantly—through communication.

Is that not an utopia far from reality? How can you imagine that? Some examples should be enough to explain the principle.

---

[26]See, for example, Chris Arnade's blog 'Why Trump voters are not "complete idiots"' (Arnade 2016).

4.4.1   Objective and Subjective Consumer Price Index

When the new currency of the euro was introduced in 2002, many citizens felt that prices had risen sharply, as services, restaurants and retailers took the opportunity of this particular moment to make higher levels of profit by using an incorrect conversion rate. In Germany, magazines and newspapers took up the widespread impression among the population and reinforced that something was wrong with the prices. 'Teuro' is a word, created in the German-speaking world, combining the German word for expensive (*teuer*) and *Euro*, which was chosen as the word of the year by the Society for German Language in 2002.

The anger about the high prices was taken up and strengthened by the media[27] and became a politically relevant problem. The official statistical inflation rate was felt to be inaccurate, unrealistic, and even politically biased in this context: '*Given the mismatch between such buying experiences and the official rate of inflation, the much-vaunted habituation to the new currency becomes a mere formula of desire. On Friday, the Federal Statistics Office announced for May a rate of inflation of 1.2 percent. In the same week Günther Hörmann of the consumer center Hamburg expected however, that a family, which needs much food, could come on an inflation rate of 15 per cent*'. [28]

Although the Federal Statistical Office commissioned a scientific study[29] to investigate and fully uncover the causes of the discrepancy between the objective and the perceived average values of the price change, statistical studies were placed in a very defensive position, which this scientific study alone did not help to ease. Even a broad initiative for communication and education could not change the sentiment once created and could not clear up reservations, prejudices and lack of statistical knowledge. The loss of confidence in the price index in particular, and official statistics in general, was considerable. Even today, so many years later, when talking with Germans, one still finds the firm assumption that the inflation rate is a politically determined number.

In terms of the perspective and goal of the greatest possible participation of civil society, a question arises as to what possible measures could be taken to avoid or at least reduce the false impression of political manipulation. Would it be possible and sensible to calculate and publish a subjective price index parallel to the official consumer price index, based on the different weighting of goods in the shopping basket (e.g. higher weight for 'out-of-pocket purchases')? Moreover, would it possibly be useful to involve citizens who are actively observing prices by providing them with a digital platform to upload the data they are gathering?

Transparency and participation of this kind could have the potential to reduce the perceived distance and also to allow people to learn more about the official index

---

[27] See, for example, http://www.teuro.de/focus/focus.html.

[28] See "Dem Teuro auf der Spur" FOCUS Magazin I Nr. 22 (2002) https://www.focus.de/finanzen/news/wirtschaft-dem-teuro-auf-der-spur_aid_203686.html.

[29] Brachinger, *Der Euro als Teuro? Die wahrgenommene Inflation in Deutschland* (Brachinger 2005).

and its methodology. Instead of requiring blind faith in statistics, this would build trust based on experience and evidence.

4.4.2 Co-production of Statistics—Participatory Data

The potential of 'Big Data' arising from any possible sources is examined by official statistics. These data are generated for specific purposes or result from technical processes. In any case, the information content for statistics must first be distilled from the dataset. In this context, approaches and ideas from the field of citizen science,[30] which aim at an active participation of volunteers in the collection of data, should be further examined.

This form of participatory filing and sharing of data and knowledge has gained momentum, especially in the areas of environment and Sustainable Development (König 2018a; Fritz et al. 2019). For example, the homepage of WeObserve states: *"WeObserve is a Coordination and Support Action which tackles three key challenges that Citizens Observatories (COs) face: awareness, acceptability and sustainability. The project aims to improve the coordination between existing COs and related regional, European and international activities. The WeObserve mission is to create a sustainable ecosystem of COs that can systematically address these identified challenges and help to move citizen science into the mainstream".*[31]

4.4.3 Participation in Indicator Design

In 2010, the UK Statistics Service was commissioned to develop and publish a set of National Statistics to understand and monitor well-being. After the programme was launched with a national debate on 'What matters to you?', to improve understanding of what should be included in measures of the nation's well-being, and after a discussion paper had summarised the output of this phase, an online consultation[32] was opened up to the wider public. This sought views on a proposed set of domains (aspects of national well-being) and headline indicators. The online consultation was open for participation between November 2010 and January 2011.

One of the challenges of such a process is to communicate in a plausible manner that there are 'participatory parts' and more 'technical parts'. Nevertheless, such a public and open consultation can make an additional contribution to bringing the

---

[30] 1. Principle of citizen science *Citizen science projects actively involve citizens in scientific endeavour that generates new knowledge or understanding.* https://ecsa.citizen-science.net/sites/default/files/ecsa_ten_principles_of_citizen_science.pdf.

[31] See https://www.weobserve.eu/ (WeObserve 2018).

[32] http://webarchive.nationalarchives.gov.uk/20120104115644, http://www.ons.gov.uk/ons/about-ons/consultations/open-consultations/measuring-national-well-being/index.html.

design of new indicators out of the sphere of experts and insiders by informing citizens as early as possible and taking account of their opinions.

However, one must consider that consultation fatigue may arise among the addressees. A consultation by scientific experts in the field of co-design[33] is therefore necessary for the success of such a project.

4.4.4 Market Research

In order to constantly develop the quality of indicators and other statistical products, it is necessary to obtain the most precise information possible about their use, misuse or non-use.[34] The application of professional methods of market research should provide evidence that is important for the product design of the future.

Figure 4.1 gives an overview of the various approaches for strengthening the involvement of users and their interests. Although many of these approaches are not completely new, as a whole they take another step in the direction of placing official statistical analysis as a service to democracy and the people. The proposals outlined here would certainly encounter difficulties in practice. For example, an online consultation costs a great deal of money and also time, both of which may be scarce in concrete situations. The co-production of data requires trust and mutual knowledge as well as corresponding IT tools; these too may be in short supply. Overall, however, it is important to overcome these hurdles and difficulties with the aim of maintaining confidence in official statistics under more difficult conditions. The cooperation between statisticians and communication experts that has successfully

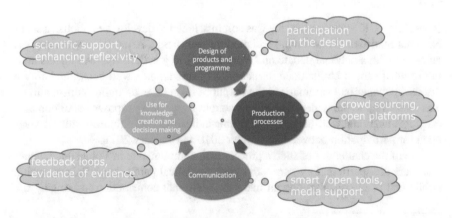

Fig. 4.1 Mainstreaming communication in the process chain

---

[33]Joost and Unteidig (2015), Gericke et al. (2018), Hisschemöller and Cuppen (2015).

[34]Lehtonen, *The multiple roles of sustainability indicators in informational governance: Between intended use and unanticipated influence* (Lehtonen et al. 2016).

developed over the past few years should therefore be complemented by cooperation with researchers in the areas of citizen science and human-centred co-design processes.[35]

4.5 Governance 4.0—Preparing for New Opportunities and Risks

Trust in statistics has always been the main objective and asset of official statistics. As stated in the previous chapters, there have been risks and incidents in past historical episodes of official statistics that have jeopardised trust, such as:

> … interference of the governments in the appointment of the staff or in the assignment of the budget, lack of a clear differentiation between official and private statistics by the public at large, diverging or contradictory data published by different government agencies on similar variables, or erroneous interpretations of statistics by the media. …But society is changing dramatically, new technologies are developing very quickly, and as a result we are now witnessing the appearance of a number of new risks, new challenges requiring new answers and solutions. Among them can be mentioned: The trust of the population in their governments and in the official institutions in general is rapidly decaying, and this lack of trust is naturally extended to the producers of official statistics. Democratic practices are being lately exerted with a rather lighter attitude than they traditionally were. … The role of statistics as an instrument for democracy is losing relevance. The changes originated by the new technological developments and the new systems of organizing work are particularly difficult to capture, and require new and imaginative methods to be adequately measured. (Martín-Guzman 2018)

Building on statistical governance, which has proven to be best practice internationally in recent years, new approaches are being sought that can provide adequate responses to new opportunities and risks.[36]

4.5.1 What Does Governance Mean?

The term 'governance' belongs to a group of terms that combines two qualities: firstly, they are poorly defined and, secondly, they have become embedded in our language at great speed.[37] This may seem paradoxical; however, it is also possible that these two qualities complement each other. It is difficult to find another technical term that could encompass legal regulations and commitments, ethical principles and other forms of policies or political commitments.

---

[35]For example Jasanoff (2003).

[36]See also J. Gray 'Quand les mondes de données sont redistribués: Open Data, infrastructures de données et démocratie' (Gray 2017).

[37]For a more detailed introduction of governance, see Chap. 3.

Two basic interpretations are covered here,[38] both of which are of interest for an analysis of statistical governance:

- governance as governing through information with a purpose to guide, steer, control or manage sectors or facets of societies,
- governance as institutional framework, in particular changes in institutions caused by the massive increase of information processes.

Governance in the area of official statistics initially includes the statistical laws at national level and the European Statistical Regulation. However, it also includes the ES Code of Practice of European Statistics and, for example, the decision of the United Nations Economic and Social Council concerning the responsibilities in the area of sustainability indicators (ECOSOC 2017). In short, governance is used here as a collective term that leaves the specific design relatively open in terms of individual governance rules.[39]

As mentioned earlier, there is a concrete reason to deal with the existing regulations. 'Official Statistics 4.0' is not only gradually different from the practice of previous periods during the last two hundred years. Due to the interlocking, manifold new factors influencing the information and knowledge society, we are confronted with structurally different working conditions, which—also beyond official statistics—require referral to 'informational governance' (Soma et al. 2016b).

4.5.2 Achieving Goals and Preventing Risks

Trust is the main and overarching goal of statistical governance. Once trust in official statistics is lost, it can take years or even decades to rebuild it (HMTreasury 1998; Thomas 2007; Sangolt 2010b). To sustain the capacity of the statistical authority to provide trustworthy and relevant statistical information is therefore the bottom line of reviews or revisions of existing statistical governance for future improvements. It is also clear that the issue of trust in the field of statistics goes far beyond the issue of trust in statistics; ultimately, trustworthy statistics (alongside fundamental rights or civil liberties) are necessary for a society itself to be trustworthy; this is based on trust being at the core of the social contract underpinning human society.

The following sub-goals serve to ensure lasting confidence:

Independence, Strength, Innovativeness: In recent years, a great deal of progress has been made to support the professional independence of official statistics, triggered by serious crises. However, as a result of the financial crisis, this progress has gone hand-in-hand with massive and permanent cuts in budgets, deterioration in staffing levels, etc. Independence combined with weakened performance cannot be a recipe for success. Moreover, it is currently vital that statistics should not be frozen in the programmes and processes of the last decades. Renewal and further development of

---

[38] See Soma et al. (2016b, p. 132).

[39] A critical reflection on governance can be found in: Brown (2015, p. 122).

the portfolio of products and services must be possible. That is why forward-looking, comprehensive programming of the statistics portfolio is crucial. It is important that a statistical service, its staff and above all its leadership have this innovative spirit and encourage and demand creativity. However, if working conditions (whether in the form of tight budgets, lack of flexibility in the programme or bureaucratic hurdles) are permanently hostile to innovation, official statistics become an archive of information on past issues; insufficient relevance and timeliness worsen the service's reputation, which results in further budget cuts, etc., leading to a vicious cycle.

Democratic participation (design process) and control (of the execution): Public administration must serve the interests of citizens and provide an equivalent of being financed with tax money. For this to happen, governmental orders provide 'checks and balances', which subject the executive to control by democratically elected bodies or other institutions (courts, courts of auditors). On the one hand, professional independence, as introduced by official statistics (in the European Union), cannot mean that a statistical office is exempted from these controls. Such absolute independence would not be democratically legitimised and could therefore be dangerous. On the other hand, specialist supervision, as is usually practiced by the political level (ministries) in relation to the administration, is incompatible in the case of statistics with the instructions of the head of the relevant office. To prevent this contradiction from creating a vacuum, different solutions have found their way into the statistical legislation of various states. At EU level, a number of special arrangements have been introduced with this objective, which has institutionalised the European Statistical Governance Advisory Board ESGAB (2010, 2016) supervisory board on statistical governance in addition to the judicial review and the European Court of Auditors. The answer to the question of the position of Eurostat as Directorate-General (ESTAT) of the European Commission also follows from the corresponding Commission decision of 2012 (European Commission 2012) and the (published) arrangements for working relations between Eurostat and the responsible Commissioner (Eurostat 2015c). Finally, it should be stressed that the European Statistical Programme is regulated by law (European Union 2011), both in terms of the planning process and its individual elements.

National, supranational and international governance must be consistent: Official statistics are less than ever before a purely national undertaking. For all important components of the statistics programme there are international standards, which to a large extent are incorporated in European statistics legislation. Revisions to these standards are extremely time-consuming and tedious processes, but their results ultimately determine how national statistics work. In general, the aspect of international comparability has become enormously important in the globalised world; this is shown by global initiatives such as the Sustainable Development Indicators (United Nations 2018), the G20 Data Gaps Initiative (IMF 2017) or the International Comparison Programme (WorldBank 2017), but above all by developments in the EU. In this respect, it is necessary to explicitly include these links and cross-relationships in any analysis and determination of governance.

4.5.3   Tailored Statistical Governance

For the very broad range of statistical products and their possible uses for information needs or decisions, there can be no uniform patent solution for the question of how statistical governance should be composed. However, it becomes clear from the previous sections that the risks to the independence of the statistics and the quality of the information depend very much on whether and to what extent they have political or financial consequences.

The more concrete policy decisions are directly linked to indicators and other statistical information, the greater the danger and temptation to influence such measures. Political interference can occur in different forms[40] and stages of the production, in the design phase, during production or dissemination. This scenario is further complicated if it is not a unilateral relationship between one statistical producer and one user, but if several statistical institutions create information of the same type in parallel, which is supposed to be combined to a consistent overall picture. In such multilateral settings it is imperative, firstly, to standardise statistical methods to a high degree and, secondly, to equip an independent institution with the competence and authority of control or audit.

This is the description of the situation in European statistics where Eurostat is legally equipped with supranational competences. Nevertheless, it remains to be decided on a case-by-case basis how far this competence should extend, i.e. how far a central influence should go, which could run counter to the needs of decentralised adaptability in such a broad statistical system as that of Europe. In particular, it is the macro-economic indicators for which the comparability of statistical results is so important that confidence in Member States' compliance with political agreements depends on it. According to political decisions, it should be 'objectively' measured and assessed whether, for example, the limits for public debt are respected. To ensure this, a fully harmonised statistical production in all Member States as well as audit-like testing and certification by Eurostat is required. This is the status quo in the area of 'Excessive Deficit Procedure (EDP)', which, however, only emerged from many rounds and years of political negotiation and experience with crises.[41] The scope and detail of the methodological standards in this particular area shows that extremely high comparability cannot be achieved without a price in other aspects of product quality: Especially the adaptability and flexibility to the constantly changing reality suffer from a methodology that is fixed by detailed laws and standards. In this respect, the aim of governance must be to ensure the optimal quality profile for the respective statistical field by taking appropriate measures.

At the global, international level, there is hardly any comparable construct to date, even if a first step in this direction has been taken with the specifications agreed in December 2018 in Katowice[42] for the measurement of information with relevance for global and national climate policy.

---

[40] See Seltzer (1994).

[41] See the "Manual on Government Deficit and Debt" (Eurostat 2016).

[42] See https://unfccc.int/node/184700.

4.5.4 Achievements of the Past 20 Years

The efforts of the past two decades have mainly focused on the professional independence of statisticians and statistical institutions.

- The community of statisticians has agreed on a Declaration of Professional Ethics, which *"consists of a statement of Shared Professional Values and a set of Ethical Principles that derive from these values"* (ISI 2010). The individual professional statistician is the focus of the declaration, aiming at giving orientation and protection by setting professional standards.
- Different codes of conduct have been developed for statistical institutes and authorities. The most influential and politically important ones are the UN Fundamental Principles of Official Statistics (United Nations 2014) and the European Statistics Code of Practice (Eurostat 2011). The latter is embedded in European legislation, such as the Treaties (European Union 2012: Art 338) and the European Statistics Regulation 223 (European Union 2015). The ES Code of Practice contains three groups of (in total) 16 principles, which address the institutional environment, the statistical process and the statistical output. A number of indicators is related to each of these principles, in order to allow for internal assessment or external reviews (since conducted twice for all EU Member States). This indicates that the ES Code of Practice is not just a bible, but rather a cornerstone of a total quality management approach in European Statistics (Radermacher 2016). Nevertheless, it is important to note that the ES Code of Practice has served as a starting point and template for the development of various codes in other regions of the world.

4.5.5 Five Pillars of Statistical Governance

In this section, a checklist of points reflects the main aspects of relevance to the governance of official statistics. The extent to which these issues manifest themselves in legal provisions or in other types of commitments (political decisions, standards, guidelines, etc.) depends on the particular context (for example, national administrative practice); therefore, these points cannot be generalised as a model solution to this puzzle. However, experience in European statistics over the past years has shown that the points in this checklist are those that collectively determine a type of solution space. Moreover, it is important not to approach the topic with the idea that simply a regulation on paper would be sufficient.

It should be considered from the very start how governance could be put into practice, what safeguards should be built in, what learning processes are foreseen, etc.

4.5.5.1   Who: Actors and Roles—Pillar One

The producers and stakeholders of official statistics participate in the process of design, information gathering and communication in different ways. Analysing their roles and functions and defining their rights and duties are the tasks of governance rules.

Producers

- Central importance in the governance belongs to the position/person at the top level of official statistics. The relevant functions and roles may possibly be merged within one position/person or may be separated by means of unmistakable mandates.

  - Statistical authority: political responsibility and accountability, setting of statistical standards and quality standards, responsibility for the programme and budget
  - Director-General/President: executive responsibility for the production and communication process and for the institution of the Statistical Office, thus in charge of allocation of the budget and of all questions concerning staff
  - Chief Statistician: coordinator of the (national) statistical system, responsibility and administrative competence to overcome resistance/reluctance by other manufacturers in the system.

- Above all, however, the question arises as to where the statistical leader is placed in the political ecosystem in terms of reputation, salary, official rank, title, direct access to the political level, etc. Likewise, the way in which recruitment and, if necessary, the dismissal of duties are regulated plays an essential role for the person/position of the President of the Statistical Office; the EU Statistics Regulation 223 lays down specific criteria for this (European Union 2015).
- A statistical system is composed of a broad set of collaborating partners with a differentiated set of tasks and responsibilities, which need to be addressed by well-balanced governance rules:

  - National producers of official statistics: national statistical system, degree of autonomy and delegated 'authority', role of official statistics of central banks
  - Science: nature and intensity of cooperation, access to aggregated and micro-data
  - Data producers: nature and intensity of cooperation, access to aggregated and micro-data
  - Media: nature and intensity of cooperation, access to aggregated and micro-data.

Interest groups, stakeholders

- Respondents: rights and obligations

  - Statistical confidentiality, privacy of information
  - (Legal) obligation to respond; a double-edged sword

– Priority for use of existing information versus survey.

- Users

  – Information needs, administrative and communicative processes
  – User classification
  – Access to statistical information and statistical (micro) data: subject to user classification
  – Civil society—advocacy role
  – Science, access to anonymised data, improvement of 'statistical literacy'
  – Media, new forms of communication, simultaneous access to timely/market-relevant indicators
  – Political participation, decision and control.

Institutions and their role in governance—participation, decision, control

- Parliament: ultimate client and addressee of official statistics products and services, legislator for statistical legal standards, programme and budget
- Statistical supervisory bodies and user committees: function to be specified in the main processes (see 'What') of official statistics
- Government: administrative responsibility in terms of budget, staff, organisation, etc., legislative initiative, essential user
- Audit authorities (e.g. audit offices): audit of the orderliness and efficiency of production processes
- Courts: case law regarding the application of statistical law
- Civil society: 'watchdog' and participating in design and communication processes as actively as possible.

4.5.5.2 What: Statistical Programme and Products, Services—Pillar Two

For the public information infrastructure provided by official statistics, it is of great importance that the range of products and services reflect both value for taxpayer money and the general information needs of society. The job of the statistics services and the budget that they receive for it must fit together.

- Design of statistical methods, products and programme, including stakeholder participation
- Planning cycles, administrative roles and responsibilities, development and negotiation of programme and budget
- Decision (binding definition), different levels of standardisation with appropriate democratic participation
- Accountability, transparency and control of execution and achievement of programme and budget.

4.5.5.3 How: Quality Assurance—Pillar Three

The quality of statistical information depends on how it is produced. Comprehensive quality management should not only ensure compliance with a high-quality standard, but also communicate transparently and certify comprehensibly.

- Institutional, administrative and legal rules prefer modern, efficient and innovative statistics production
- Institutional, administrative and legal rules give the Director General the necessary freedom of operation concerning the management of the statistical institution (staff, organisation, finances, technical infrastructure, etc.)
- Ethical codes, good governance principles
- State-of-the-art scientific approach and standards at all levels
- Cooperation with and reviews by peers/counterparts in the statistical community
- Efficiency criteria, also for cooperation between statistics producers
- Quality management.

  - Quality management methods, e.g. from European Foundation for Quality Management (EFQM)
  - Quality control, proportional to the policy impact of statistics (e.g. financial statistics, public debt indicators)
  - Reporting on quality (e.g. quality reports after completion of a survey, as provided for in legal standards, Commitment on Confidence in the European Statistical System ESS)
  - Communication of quality assurance, branding (for example quality declaration by the ESS), labelling of quality profiles for statistical products.

4.5.5.4 Confidentiality, Data Ownership, Access to Data—Pillar Four

In the information society, data has great economic value. As a result, ownership issues, rights of use and obligations to provide or keep confidential information are of paramount importance to informational governance. With new developments in the field of public administration, new tasks in data management will be created and competencies and responsibilities will have to be regulated. Official statistics have a lot to offer for these new tasks. However, administrative and statistical work areas must be separated from each other in such a way that confidentiality of the statistical micro-data is ensured. Implementing shared EU rules on access to data from non-public sources for statistical purposes will allow the ESS to play its role successfully in current and future difficult times.

- Legal basis for collection or (re-)use of individual data

- Statistical confidentiality and general data protection,[43] ensuring consistency between two legal concepts; principles and guidelines concerning statistical confidentiality and data protection
- Role and function of official statistics' role in the data economy and data space,[44] access to (public) statistical data (government-to-business [G2B]), sharing private sector data (business-to-government [B2G])[45]
- Special legal provisions for the relationship between administrative and statistical registers, especially if both are operated by official statistics services
- Special legal provisions for data linkage by statistical institutions.[46]

4.5.5.5 International Cooperation and Relations, Standards, Governance—Pillar Five

With increasing cooperation between the statistical authorities of different states, especially within the framework of European statistics, cross-border regulations, supranational and international agreements are becoming increasingly important. The difficult negotiations over the past few years for the exchange of confidential data as part of the reform of intra-trade[47] statistics are one of many examples. Many factors have to be balanced in such regulations: the change in the political framework conditions, and in particular further developments of the political governance of the EU, must be considered.

- European statistics: In recent years, a wide range of rules have been laid down, through which the official statistics of the Member States have been integrated into the network of partnership relations with each other and with Eurostat in a statutory system. At the moment, it is completely unclear in which direction the European project will develop. But as the example of the UK's exit ('Brexit') demonstrates, one thing is clear: any form of Europe needs solid statistics, even in the preparation and discussion of possible political alternatives. Providing adequate governance as an institutional framework is an urgent task to be undertaken along this path.
- UN, Organisation for Economic Co-operation and Development (OECD), United Nations Economic Commission for Europe (UNECE): As an example, reference

---

[43]For data protection in the EU see https://ec.europa.eu/info/law/law-topic/data-protection/data-protection-eu_en.

[44]EU policy: https://ec.europa.eu/digital-single-market/en/policies/building-european-data-economy.

[45]*Synopsis report of the public consultation on the revision of the Directive on the reuse of public sector information* https://ec.europa.eu/digital-single-market/en/news/synopsis-report-public-consultation-revision-directive-reuse-public-sector-information.

[46]See the report 'Joining up Data for Better Statistics' by the UK Statistics Authority (Office for Statistics Regulation 2018).

[47]See the SIMSTAT approach in the revised regulation for EU business statistics (Eurostat 2015b).

should be made to global cooperation in the development of sustainability indicators, which does not only touch on statistical–methodological problems, but also on the distribution of roles, responsibilities and other governance issues.

- International Monetary Fund (IMF), World Bank and UN agencies: As a rule, these areas will be more concerned with the technical aspects of individual statistical areas. Nevertheless, issues of cooperation, data flows, etc. usually play a role.

4.5.6 The Data-Information-Knowledge Nexus and Official Statistics

As an adequate approach for reacting to the rapidly transforming political landscape caused by the digital revolution, globalisation, the crisis of the nation state and the changing position of science in society, Soma et al. (2016b, p. 134) widen the concept of governance in order to introduce an 'informational governance',[48] outlining this "*along four interrelated themes:*

- *Processes of information construction: how 'governing through information' appears and influences institutional change,*
- *Information processing through new technology, for example, social media: how information construction through use of new technology affects diversification of future governance arrangements,*
- *Qualities of transparency and accountability: how "governing through information" appears and influences institutional change,*
- *Fourth, institutional change: how new institutional arrangements for governing are emerging in the Information Age as a matter of new ICT developments, globalization, as well as new roles of state and science.*"

They further state that: "*growing uncertainties and complexities are partly caused by difficulties in controlling information flows in the more globalised world. Because the state and science increasingly are lacking the authority to unilaterally solve controversies bound up with politics and struggles on knowledge claims, problems of definitions, trust and power are increasing*" (Soma et al. 2016b, p. 133).

Central to all considerations for statistical governance is the entirely reviewed role of citizens (Soma et al. 2016a).

In essence, the situation of official statistics will continue to be determined by techniques (tools), ethics (patterns of behaviour) and politics (questions of institutional set-up or communication). However, in rapidly changing circumstances, it is

---

[48] 'The concept of informational governance has emerged to capture these new challenges of environmental governance in the context of the Information Age. The logic of informational governance stems from the observation that information is not only a source for environmental governance arrangements, but also that it contributes to transformation of environmental governance institutions. Such societal transformation refers to how the raise of information technology, flows and networks leads to a fundamental restructuring of governance processes, structures, practices and power relations' (Soma et al. 2016b, p. 131).

important that official statistics services play their important social role by adequately adapting the rules, principles and resources that shape their working conditions. They should be enabled to act proactively in the sense of educating liberal democratic societies (Stapleford 2015).

"In January 2018, the European Commission set up a high-level group of experts ("the HLEG") to advise on policy initiatives to counter fake news and disinformation spread online. The main deliverable of the HLEG was a report designed to review best practices in the light of fundamental principles, and suitable responses stemming from such principles" (HLEG 2018). A review of statistical governance requires a broad approach, similar to that chosen here for the media. Stakeholders of official statistics should jointly define a multidimensional approach to address the issues ahead and constructively continue the successfully established governance of statistics.

In the rapidly evolving variety of data science disciplines, there is a considerable risk that existing knowledge and evolved structures will go unused in governance and that the wheel will be reinvented many times. Essentially, the goal of governance is to create and maintain trust in information and, where lost, to regain it. But if such fundamental principles are neglected by other information producers and if unrealistic expectations are created,[49] then it is difficult for users to recognise differences in quality.

The question therefore arises as to whether the statistical governance structures dating back to the discussions of the 1990s,[50] which focused on nationally organised public administrations of statistics with their risks, still meet today's challenges. Rather, it seems sensible and necessary to subject these fundamental principles to a review and revision process.

Official statistics services are therefore required to take the initiative here, to contribute their knowledge and to play an active, coordinating and integrating role in the discussion between different disciplines.

Questions that need to be addressed and answered (amongst others) are:

- Informational governance for the data sciences: lessons to be learnt and transferred from statistics
- Statistics and data science in public administration: who is responsible for what?
- Professional values and ethics, revision of the European Statistical System Code of Practice, evaluation of the status quo, analysis, gaps, recommendations, ethics for all three core statistical processes (design, production, communication)
- Governance for different types of statistical products (indicators 'with authority'; indicators, accounts, statistics and their quality profiles; experimental statistics)
- Ethics for decision-makers and their scientific services
- Statistical competence: intensified cooperation between the education system (including vocational training) and official statistics services

---

[49]Wigglesworth, *Can Big Data revolutionise policymaking by governments?* (Wigglesworth 2018); Heubl, *Night light images paint accurate picture of China GDP* (Heubl 2018).

[50]See in particular William Seltzer's paper 'Politics and Statistics: Independence, Dependence, or Interaction?" (Seltzer 1994).

- Official statistics services' obligations and rights in the data economy (B2G, G2B, G2G).
- International statistical governance

  - Global conventions that go beyond today's recommendations
  - A new regulatory framework for access to privately owned data for official statistics.

4.6 Different Communities and Their Isolated Discussions

In his book "Postfaktisch" Vincent F. Hendricks presents a scale (see Fig. 4.2) of information quality *"in which true and different forms of false statements and strategies undermining truth face each other at opposite ends"* (Hendricks and Vestergaard 2018, p. 75).

While he goes into detail on the various variants of misinformation, such as distorted statements, lies, fake news, it remains unclear what he understands by 'true statements'. As a definition he offers 'Verified facts'. In the light of the considerations set out in Chap. 2, it remains to be seen whether the use of the term 'truth' is helpful and appropriate here. Nevertheless, from the statistician's point of view the question arises as to what 'facts' are and how they should and can be 'verified'.

Three aspects are decisive for the quality of statistical information: first, statistical measurement quality; second, theoretical–methodological consistency; and third, relevance for information needs and decisions. Only if all three aspects are achieved

Zone 1	
True statement	*Verified facts*
ZONE 2	
Distorted statements	*Framing, acute angles, omission, "selected facts"*
Unsubstantiated statements	*Rumors (maybe true, maybe false)*
ZONE 3	
False statements	*False rendering of facts, in contradiction to these lies*
Bullshit	*False rendering of one's own motives and goals, misrepresentation, faking, dissolution of the separation between true and false*
Fake News	*Fake news, false reproduction of motifs and goals with simulation of journalism and thus truthfulness*

Fig. 4.2 Information quality scale. Adapted from Hendricks and Vestergaard (2018, p. 76)

satisfactorily, or better 'adequately', can a statistical number, indicator, graph or map play its role: Because then it is fit for purpose.

Unfortunately, however, these three aspects are traditionally dealt with by and in different communities, which is very detrimental to the achievement of the stated objective. For example, the Science and Technology Studies (Latour et al. 1986; Jasanoff and Science 1995) located in sociology or the discourse on 'Governing by the Numbers' (Bartl et al. 2019) are little or not at all known in statistics. On the other hand, little is known in these scientific Communities about the progress that statisticians have made in recent years in terms of information quality (and its verification).[51]

Especially under the current conditions of digitalisation, globalisation and the increasingly widespread scepticism towards experts and facts, it is therefore important to close the gaps between the discussions in different communities.

For this reason, it must be possible to create common and interdisciplinary platforms and channels for the exchange of information and opinions. Such a role could be played by scientific journals and conferences with such overarching topics if it were possible to bring together qualified contributions from authors from statistics (design, production, communication), sociology or political science and from a methodological–scientific subject area. This requires concrete questions and applications, preferably in the form of an important statistical project such as the restructuring of demographic statistics including the census, the Sustainable Development Indicators or the International Comparison Programme.

Such interdisciplinary cooperation has already been intensified in recent years both with data journalists and with data scientists. It would now be very important to build a bridge to the academic fields, which deal professionally with the manifold interactions between statistics and society.

Parallel to this interdisciplinary cooperation in research and development, a similar extension and improvement of training in statistics is also necessary. While it will hardly be possible to recruit graduates from tailor-made education and study courses for the wide variety of professional qualifications that are needed today in official statistics and which are combined there, more could and should be done to prepare the next generation of statisticians adequately for their important work in statistical institutions.

Statistical offices have for some time offered programmes for the vocational education and training of young statisticians.[52] With the 'European Master in Official Statistics (EMOS)',[53] this has been continued to the extent that a cooperation with a large number of interested universities has been institutionalised with the aim of offering a degree that—with the help of a special curriculum—prepares for deployment

---

[51] See for example the Quality Framework of European Statistics here https://ec.europa.eu/eurostat/web/quality/european-statistics-code-of-practice.

[52] See, for example, the European Statistical Trainings Programme (ESTP) here https://ec.europa.eu/eurostat/cros/content/european-statistical-tranings-programme-estp_en.

[53] https://ec.europa.eu/eurostat/web/european-statistical-system/emos.

in official statistics. For these training courses and degrees[54] it would be desirable in the future to increasingly integrate those aspects that prepare for the role of statistics in the new informational ecosystem around the Internet of Things, not least involving the sociological disciplines mentioned above. However, due to the wide variety of topics that should be covered in such a programme as a whole, it is likely to be necessary and useful to include optional subjects and topics that participating universities can focus on in addition to a core programme of compulsory learning content. In this way it would be possible, for example, to provide for specialisation in National Accounts as an option and thus close gaps in the current EMOS programme.

References

Arnade, Chris. 2016. *Why Trump Voters are not "Complete Idiots"*. Accessed May 31, 2016. https://medium.com/@Chris_arnade/trump-politics-and-option-pricing-or-why-trump-voters-are-not-idiots-1e364a4ed940#.faldoe9vg.

Bartl, Walter, Christian Papilloud, and Audrey Terracher-Lipinsky. 2019. Governing by Numbers—Key Indicators and the Politics of Expectations. *Historical Social Reasearch* 44: 1–339.

Bean, Charles. 2016. *Independent Review of UK Economic Statistics*, London, UK Government.

Beck, Ulrich. 1998. *Risk Society Towards a New Modernity*. London: Sage.

Benessia, A., S. Funtowicz, M. Giampietro, A. Guimaraes Pereira, J. Ravetz, A. Saltelli, R. Strand, and J.P. van der Sluijs. 2016. *The Rightful Place of Science: Science on the Verge*. Tempe, AZ: Consortium for Science, Policy and Outcomes.

Brachinger, H.W. 2005. Der Euro als Teuro? Die wahrgenommene Inflation in Deutschland. *Wirtschaft und Statistik*.

Braungardt, Jürgen. 2018. *Immanuel Kant: What is Enlightenment? (1784)*. Braungardt, Jürgen. Accessed April 20, 2018. http://braungardt.trialectics.com/philosophy/early-modern-philosophy-16th-18th-century-europe/kant/enlightenment/.

Brown, W. 2015. *Undoing the Demos: Neoliberalism's Stealth Revolution*. Cambridge, MA: MIT Press.

Cao, Longbing. 2017a. Data Science: A Comprehensive Overview. *ACM Computing Surveys* 50: 1–42.

Cao, Longbing. 2017b. Data Science: Challenges and Directions. *Communications of the ACM* 80: 59–68.

Dasgupta, Rana. 2018. The Demise of the Nation State. *The Guardian*, April 5, 2018.

Davies, William. 2016. *The Limits of Neoliberalism—Authority, Sovereignty and the Logic of Competition*. London: SAGE Publications.

Davies, William. 2017. How Statistics Lost Their Power—and Why We Should Fear What Comes Next. *The Guardian*.

De Clerck, J-P 2017. What is the Internet of Things? Internet of Things Definitions and Segments. In *i-SCOOP*.

Desrosières, Alain. 1998. *The Politics of Large Numbers—A History of Statistical Reasoning*. Cambridge, MA: Harvard University Press.

Diakopoulos, Nicholas. 2015. Accountability in Algorithmic Decision-making. *Communications of the ACM* 59: 56–62.

Diaz-Bone, Rainer, and Emmanuel Didier (ed.). 2016. *Conventions and Quantification—Transdisciplinary Perspectives on Statistics and Classifications*.

---

[54] See the current learning outcomes of EMOS here https://ec.europa.eu/eurostat/cros/system/files/emos_learning_outcomes_2018.pdf.

ECOSOC. 2017. *ECOSOC Adopts SDG Indicator Framework.* Accessed June 7, 2017. http://sdg. iisd.org/news/ecosoc-adopts-sdg-indicator-framework/.

ECSA. 2016. *ECSA Policy Paper #3 Citizen Science as part of EU Policy Delivery—EU Directives,* 4. Berlin: ECSA.

ESGAB. 2010. *Second Annual Report to the European Parliament and the Council on the Implementation of the European Statistics Code of Practice by Eurostat and the European Statistical System as a Whole.* European Statistical Governance Advisory Board.

ESGAB. 2016. *ESGAB Annual Report 2016.* Luxembourg: European System Governance Advisory Board.

ESRG. 2016. *Report of the Economic Statistics Review Group (ESRG).* Dublin: CSO Ireland.

European Alpbach Forum. 2016. *An Introduction to "New Enlightenment".* https://www.alpbach. org/en/forum2016/programme-2016/new-enlightenment-an-introduction-by-the-presidents-% 20of-the-european-forum-alpbach/.

European Commission. 2012. *Reinforcing Eurostat, Reinforcing High Quality Statistics,* ed. European Commission. Brussels: European Commission.

European Statistical System Committee. 2018. Bucharest Memorandum on Official Statistics in a Datafied Society (Trusted Smart Statistics). In *DGINS 2018,* ed. Eurostat. Bucharest: Statistics Romania.

European Strategy and Policy Analysis System. 2019. *Global Trends to 20130—Challenges and Choices for Europe,* 42. Brussels: European Strategy and Policy Analysis System ESPAS.

European Union. 2011. European Statistical Programme 2013–2017. In *COM(2011) 928 Final,* ed. The European Parliament and the Council. Brussels: European Commission.

European Union. 2012. The Treaty on the Functioning of the European Union (TFEU), ed. European Commission. Brussels: Official Journal of the European Union C326/193.

European Union. 2015. Regulation (EC) No 223/2009 of the European Parliament and of the Council of 11 March 2009 on European statistics and repealing Regulation (EC, Euratom) No 1101/2008 of the European Parliament and of the Council on the transmission of data subject to statistical confidentiality to the Statistical Office of the European Communities, Council Regulation (EC) No 322/97 on Community Statistics, and Council Decision 89/382/EEC, Euratom establishing a Committee on the Statistical Programmes of the European Communities. In *2009R0223—EN— 08.06.2015—001.001—1,* ed. European Union. Luxembourg: © European Union, https://eur-lex. europa.eu, 1998–2019.

Eurostat. 2011. *European Statistics Code of Practice for the National and Community Statistical Authorities—Adopted by the European Statistical System Committee 28th September 2011,* ed. Eurostat. Luxembourg: Eurostat.

Eurostat. 2013b. *European System of Accounts ESA 2010.* Luxembourg: Publications Office of the European Union.

Eurostat. 2015a. *Practical Arrangements Governing Working Relations Between Commissioner Thyssen, Her Cabinet and Eurostat.* Luxembourg: Eurostat.

Eurostat. 2015b. *Implementation of the Framework Regulation Integrating Business Statistics (FRIBS).* Eurostat. http://ec.europa.eu/eurostat/about/opportunities/consultations/fribs.

Eurostat. 2016. *Manual on Government Deficit and Debt—Implementation of ESA 2010.* Luxembourg: Publications Office of the European Union.

Eyraud, Corine. 2018. Stakeholder Involvement in the Statistical Value Chain: Bridging the Gap Between Citizens and Official Statistics. In *Power from Statistics: Data, Information and Knowledge—Outlook Report—2018 Edition,* ed. Eurostat. Luxembourg: Publication Office of the European Union.

Foucault, Michel. 1991. 'Governmentality.' in Graham Burchell, Colin Gordon and Peter Miller (eds.), *The Foucault Effect* (Chicago University Press: Chicago).

Fritz, Steffen, Linda See, Tyler Carlson, Mordechai Muki Haklay, et al. 2019. Citizen science and the United Nations Sustainable Development Goals. *Nature Sustainability* 2: 922–930.

Georgiou, Andreas V. 2018. *A New Statistical System for the European Union.* Brussels: Bruegel.

Gericke, Kilian, Boris Eisenbart, and Gregor Waltersdorfer. 2018. Staging Design Thinking for Sustainability in Practice: Guidance and Watch-Outs. In *Sustainability Science*, ed. Ariane König. New York: Routledge.

GfdS. 2016. *GfdS wählt » postfaktisch « zum Wort des Jahres 2016*, ed. Gesellschaft für Deutsche Sprache. Wiesbaden.

Global Commission on the Future of Work. 2019. *Work for a Brighter Future*, ed. International Labour Organization ILO, 71. Geneva: International Labour Organization ILO.

Gray, Jonathan. 2017. Quand les mondes de données sont redistribués: Open Data, infrastructures de données et démocratie. *Statistique et Société* 5: 29–34.

Haklay, Muki. 2015. Citizen Science and Policy: A European Perspective. In *Case Study Series*, 61. Washington, DC: Wilson Center COMMONS LAB.

Hale, Thomas, David Held, and Kevin Young. 2013. *Gridlock: Why Global Cooperation is Failing When We Need It Mos*. Cambridge: Polity Press.

Hand, D.J. 2004. *Measurement Theory and Practice: The World Through Quantification*. London: Arnold.

Hendricks, Vincent F., and Mads Vestergaard. 2018. *Postfaktisch—Die neue Wirklichkeit in Zeiten von Bullshit, Fake News und Verschwöruungstherien*. München: Karl Blessing Verlag.

Heubl, Ben. 2018. Night Light Images Paint Accurate Picture of China GDP. *NIKKEI ASIAN REVIEW*, 24 March 2018.

Hisschemöller, Matthijs, and Eefje Cuppen. 2015. Participatory Assessment: Tools for Empowering, Learning and Legitimating? In *The Tools of Policy Formulation*, ed. Andrew J. Jordan and John R. Turnpenny. Cheltenham: Edward Elgar Publishing Limited.

HLEG. 2018. "A multi-dimensional approach to disinformation - Final report of the High Level Expert Group on Fake News and Online Disinformation." In, 39. Luxembourg: European Commission.

HMTreasury. 1998. *Statistics: A Matter of Trust*. London: HM Treasury.

IMF. 2017. *Prinicple Global Indicators*. IMF. http://www.principalglobalindicators.org/?sk= E30FAADE-77D0-4F8E-953C-C48DD9D14735.

Independent Commission for Sustainable Equality. 2018. *Sustainable Equality—Well-Being for Everyone in a Sustainable Europe*, ed. Marcel Mersch, 193. Brussels: Group of the Progressive Alliance of Socialists & Democrats in the European Parliament.

Intergovernmental Science-Policy Platform on Biodiversity and Ecosystem Services. 2019. *Summary for Policymakers of the Global Assessment Report on Biodiversity and Ecosystem Services*, 39. Bonn, Germany: Intergovernmental Science-Policy Platform on Biodiversity and Ecosystem Services (IPBES).

ISI. 2010. *ISI Declaration on Professional Ethics*. Hague: International Statistical Institute.

Jasanoff, Sheila. 2003. Technologies of Humility: Citizen Participation in Governing Science. *Minerva* 41: 223–244.

Jasanoff, Sheila. 2004. *States of Knowledge: The Co-production of Science and the Social Order*. London: Routledge.

Jasanoff, S., and Society for Social Studies of Science. 1995. *Handbook of science and Technology Studies*. Thousand Oaks: Sage Publications.

Joost, Gesche, and Andreas Unteidig. 2015. Design and Social Change: The Changing Environment of a Discipline in Flux. In *Transformation Design*, ed. Wolfgang Jonas, Sarah Zerwas, and Kristof von Anshelm. Birkhäuser: Basel.

König, Ariane (ed.). 2018. *Sustainability Science*. New York: Routledge.

Latour, B., S. Woolgar, and J. Salk. 1986. *Laboratory Life: The Construction of Scientific Facts*. Princeton: Princeton University Press.

Lehtonen, Markku. 2015. Indicators: Tools for Informing, Monitoring or Controlling? In *The Tools of Policy Formulation—Actors, Capacities, Venues and Effects*, ed. Andrew J. Jordan and John R. Turnpenny. Cheltenham: Edward Elgar Publishing.

Lehtonen, Markku, Léa Sébastien, and Tom Bauler. 2016. The Multiple Roles of Sustainability Indicators in Informational Governance: Between Intended Use and Unanticipated Influence. *Current Opinion in Environmental Sustainability* 2016: 1–9.

Lohr, Steve. 2016. Civility in the Age of Artificial Intelligence. In *ODBMS.org*. Zicari, Roberto.

Mance, Henry. 2016. Britain has had Enough of Experts, Says Gove. *Financial Times*.

Martín-Guzman, Pilar. 2018. Old and New Risks for the Credibility of Official Statistics: Comments from a User. In *Conference of European Statistical Stakeholders*, ed. University Bamberg. Bamberg, Germany: University Bamberg.

Moulton, Brent R., and Peter van de Ven. 2018. Addressing the Challenges of Globalization in National Accounts. In *Conference on Research in Income and Wealth "The Challenges of Globalization in the Measurement of National Accounts"*. Washington.

Office for Statistics Regulation. 2018. Joining Up Data for Better Statistics. In *Systemic Review Programme*, ed. UK Statistics Authority, 39. London: UK Statistics Authority.

Oxford Dictionairies. 2016. *Oxford Dictionairies Word of the Year 2016*. Accessed June 26, 2018. https://www.oxforddictionaries.com/press/news/2016/12/11/WOTY-16.

Peruzzi, Alberto. 2017. Complexity: Between Rhetoric and Science. In *Complexity in Society: From Indicators Construction to their Synthesis*, ed. F. Maggino. Cham: Springer International Publishing.

Pope Francis. 2014. *Address to European Parliament*, Strasbourg, France, November 25, 2014.

Radermacher, Walter. 2016. Quality Declaration of the European Statistical System—Inauguration. In *CESS 2016*. Budapest.

Radermacher, Walter J., and Emanuele Baldacci. 2016. Official Statistics for Democratic Societies— Dinner Speech at the CESS 2016, Budapest. In *Conference of European Statistical Stakeholders*. Budapest.

Randers, Jorgen, Johan Rockström, Per Espen Stoknes, Ulrich Golücke, David Collste, and Sarah Cornell. 2018. *Transformation is Feasible—How to Achieve the Sustainable Development Goals Within Planetary Boundaries—A Report to the Club of Rome, for its 50 years Anniversary 17 October 2018*. Stockholm: Stockholm Resilience Centre, Stockholm University, Norwegian Business School, Global Challenges Foundation.

Ricciato, Fabio, Michail Skaliotis, Albrecht Wirthmann, Kostas Giannakouris, and Fernando Reis. 2018. Towards a Reference Architecture for Trusted Smart Statistics. In *DGINS 2018*. Bucharest: Statistics Romania.

Roser, Max. 2018. Most of Us are Wrong About How the World has Changed (Especially Those Who are Pessimistic About the Future). In *Our World In Data*.

Ryan, Liz. 2014. 'If You Can't Measure It, You Can't Manage It': Not True. In *Forbes/Leadership*. Forbes.

Sangolt, Linda. 2010a. A Century of Quantification and "Cold Calculation". Trends in the Pursuit of Efficiency, Growth and Pre-eminence. In *Between Elightenment and Disaster—Dimensions of the Political Use of Knowledge*, ed. Linda Sangolt. Brussels: P.I.E. Peter Lang.

Sangolt, Linda. 2010b. *Between Enlightenment and Disaster: Dimensions of the Political Use of Knowledge*. Brussels: P.I.E. Peter Lang.

Seltzer, William. 1994. *Politics and Statistics: Independence, Dependence, or Interaction?*, ed. UN Dept. of Econ. and Soc. Information and Policy Analysis. New York: UN.

Soma, Katrine, Marleen C. Onwezen, Irini E. Salverda, and Rosalie I. van Dam. 2016a. Roles of citizens in environmental governance in the Information Age—four theoretical perspectives. *Current Opinion in Environmental Sustainability* 2016: 122–130.

Soma, Katrine, Bertrum H. MacDonald, Catrien J.A.M. Termeer, and Paul Opdam. 2016b. Introduction article: informational governance and environmental sustainability. *Current Opinion in Environmental Sustainability* 2016: 131–139.

Stapel-Weber, Silke, and John Verrinder. 2016. Globalisation at Work in Statistics—Questions Arising from the 'Irish Case. *Eurona—Eurostat Review on National Accounts and Macroeconomic Indicators*, 29–44.

Stapel-Weber, Silke, Paul Konijn, John Verrinder, and Henk Nijmeijer. 2018. Meaningful Information for Domestic Economies in the Light of Globalization—Will Additional Macroeconomic Indicators and Different Presentations Shed Light? In *Conference on Research in Income and Wealth "The Challenges of Globalization in the Measurement of National Accounts"*. Washington.

Stapleford, Thomas A. 2015. Price Indexes, Political Judgments, and the Challenge of Democratic Control. In *Ottawa Group—International Working Group on Price Indices—Fourteenth Meeting*. Tokyo, Japan: Statistics Japan.

Stiglitz, Joseph E., Jean-Paul Fitoussi, and Martine Durand. 2018a. *Beyond GDP—Measuring What Counts for Economic and Social Progress*. Paris: OECD Publishing.

Stiglitz, Joseph E., Jean-Paul Fitoussi, and Martine Durand (eds.). 2018b. *For Good Measure, Advancing Research on Well-being, Metrics Beyond GDP*. Paris: OECD Publishing.

Sturgeon, Timothy J. 2013. *Global Value Chains and Economic Globalization—Towards a New Measurement Framework*. Luxembourg: Eurostat.

Thomas, Ray. 2007. *Who is in Charge of Public Statistics?*, radstats. http://www.radstats.org.uk/no094/Thomas94.pdf.

United Nations. 2014. *Fundamental Principles of Official Statistics*. New York.

United Nations. 2018. *Sustainable Development Goal indicators website*. United Nations. Accessed August 17, 2018. https://unstats.un.org/sdgs/.

Villani, Cédric. 2018. For a Meaningful Artificial Intelligence—Towards a French and European Strategy. In *Mission Assigned by the Prime Minister Édouard Philippe*, 151. Paris: AI for Humanity—French Strategy for Artificial Intelligence.

WeObserve. 2018. *An Ecosystem of Citizen Observatories for Environmental Monitoring*. WeObserve, Accessed August 17, 2018. https://www.weobserve.eu/.

Wigglesworth, Robin. 2018. Can Big Data Revolutionise Policymaking by Governments? *Financial Times*, 31 January 2018.

World Economic Forum. 2019. The Global Risks Report. In *The Global Risks Report*, ed. World Economic Forum WEF, 107. Geneva: World Economic Forum WEF.

WorldBank. 2017. *International Comparison Program (ICP)*. http://www.worldbank.org/en/programs/icp.

Chapter 5
A Confident Look into the Future
of Official Statistics

Official statistics have shown an excellent record in the role of trustworthy authority in the intersection of three fundamental rights: privacy (the right of a person to privacy), freedom of information (the right of a person to open and transparent information) and statistics (the right of a person to live in an informed society[1]).

200 years of experience and a good reputation are assets as important as the profession's stock of methods, international standards and well-established routines and partnerships. It is now important to nurture, renew and increase this capital by making new partnerships with young and creative professionals from statistics, data sciences or other disciplines, opening up new data sources and providing new products and services. Preserving the assets and heritage of the past while being innovative is the great challenge.

This study has been geared towards this goal. Essentially, two starting points and methodologies were used. First, official statistics was considered an actor and player in the information business, with a consistent application of economic categories and concepts of management. Statistics are products and services that are to be produced with excellent quality and for lasting customer satisfaction. Second, a reflective sociological approach was used to analyse the effects and side effects of official statistics as a public information infrastructure.

While the first business management approach has now found its way into the quality management systems used by many statistical offices, the second, sociological, approach remains relatively rare. However, these two points of view complement each other well; the business resides at a micro-level, while the sociological encompasses the social whole, the macro-level.

To successfully address new partnerships, data, opportunities and risks, it is vital that official statistics services and professional statisticians are aware of their strengths and weaknesses, that they know the DNA of their business and that they learn from their history. A 'reflexive form of scientization' (Beck 1998, p. 155)

---

[1]The right of a person to have access to and be a stakeholder in the provision of good quality statistical information intended to steer and govern a society.

© Springer Nature Switzerland AG 2020
W. J. Radermacher, *Official Statistics 4.0*,
https://doi.org/10.1007/978-3-030-31492-7_5

in official statistics requires considerably more than statistical methods, data sciences and information technology. Therefore, business administration and sociological knowledge should also belong to the standard building blocks of education for the professional statisticians of the future.

With such an extensive knowledge base and toolkit and teams of highly skilled professionals, official statistics will undoubtedly succeed in making the transition to the era of digitisation and globalisation in the form of 'Official Statistics 4.0'.

Reference

Beck, Ulrich. 1998. *Risk Society Towards a New Modernity*. London: Sage.

Printed in the United States
By Bookmasters